Soviet Science and Engineering in the Shadow of the Cold War

The 1950s were a vital time in the history of science. In accordance with the intensification of the Cold War, many scientific talents were mobilized to several military-related research and development projects not only in the United States but also in the Soviet Union. Contrary to the expectation of General Leslie Groves, a leader of the Manhattan Project, the Soviet Union succeeded in their nuclear weapon development in a very short time. And then, by the end of the decade, mankind reached the dawn of the Atomic Age proper with the beginning of the operation of the world's first civil nuclear power plant in Obninsk in 1954. The risky and costly developments of new weapons such as rockets, jet warplanes, and computers were achieved by the Soviet Union in a very short time after World War II in spite of the heavy economic damage caused by the battles with German troops in Soviet territory.

Why were such a great number of scientific talents mobilized to various Soviet Cold War research and development projects? What were the true natures and real consequences of the rushed Cold War projects? How did Soviet scientists approach the nuclear age? Thanks to the study of formerly classified Soviet archives, a more nuanced view of Soviet society has become possible. To resolve the above-mentioned questions, Ichikawa analyses the complicated interactions among various factors, including the indigenous contradictions in the historical development of science in the Soviet Union; conflicts among the related interest groups; relationships with the political leadership and the military; the role of ideology; and others.

Hiroshi Ichikawa is Professor of the History of Science and Technology in the Graduate School of Integrated Arts and Sciences at Hiroshima University, Japan.

Routledge Studies in the History of Science, Technology and Medicine

Spatializing the History of Ecology
Sites, Journeys, Mappings
Edited by Raf de Bont and Jens Lachmund

Cancer, Radiation Therapy and the Market
Barbara Bridgman Perkins

Science and Ideology
A Comparative History
Mark Walker

Closing the Door on Globalization: Internationalism, Nationalism, Culture and Science in the Nineteenth and Twentieth Centuries
Fernando Clara and Cláudia Ninhos

Johann Friedrich Blumenbach: Race and Natural History, 1750–1850
Edited by Nicolaas Rupke and Gerhard Lauer

Health Policies in Transnational Perspective: Europe in the Interwar Years
Josep-Lluís Barona-Vilar

Urban Histories of Science
Making Knowledge in the City, 1820–1940
Edited by Oliver Hochadel and Agustí Nieto-Galan

Pioneering Health in London, 1935–2000
The Peckham Experiment
David Kuchenbuch

Soviet Science and Engineering in the Shadow of the Cold War
Hiroshi Ichikawa

For the full list of titles in the series, please visit: www.routledge.com/Routledge-Studies-in-the-History-of-Science-Technology-and-Medicine/book-series/HISTSCI

Soviet Science and Engineering in the Shadow of the Cold War

Hiroshi Ichikawa

Routledge
Taylor & Francis Group
LONDON AND NEW YORK

First published 2019
by Routledge
2 Park Square, Milton Park, Abingdon, Oxon OX14 4RN

and by Routledge
52 Vanderbilt Avenue, New York, NY 10017

First issued in paperback 2020

Routledge is an imprint of the Taylor & Francis Group, an informa business

British Library Cataloguing-in-Publication Data
A catalogue record for this book is available from the British Library

Library of Congress Cataloging-in-Publication Data
A catalog record for this book has been requested

ISBN 13: 978-0-36-758522-8 (pbk)
ISBN 13: 978-1-138-55245-6 (hbk)

Typeset in Galliard
by Apex CoVantage, LLC

To Kaji-san

Contents

Preface

To design a future in which the achievements of science and engineering are mostly utilized for human welfare, we must overcome the structural problem of "the Cold War science and engineering" in which a tremendous amount of funds, research resources (manpower, research facilities, equipment, devices, and other resources) and productive resources have been consumed for various military and military-related projects of the states. Similar to the United States, in the Soviet Union, various state scientific and technological projects were carried out during the Cold War. This book is the fruit of the author's research on the initiation of "the Cold War of science and engineering" in the Soviet Union during the early years of the Cold War.

However, there are some chapters that are omitted from this book; it does not refer to nuclear weapon development, which must be regarded as the highpoint of the various scientific efforts during the Cold War, and other important issues such as nuclear submarine development and radar development.

Due to the lack of the possibility to access primary research sources such as archival documents and records and, moreover, due to the difficulty in providing readers with any further original arguments that cannot be found in the earlier literature on these subjects, the author omitted these chapters, although these issues were argued in a book written by the author in Japanese [Ichikawa, Hiroshi, *Reisen to Kagaku-Gijutsu: Sorenpô 1945–1955.* (*The Cold War, Science and Technology: The Soviet Union, 1945–1955*). Kyoto, Minerva Shobô, 2007] on the basis of literature, books, and papers already published in Russian or English to provide Japanese readers with an overview of the Soviet Union's scientific efforts during the Cold War. If you are interested in these subjects, please refer to the referenced literature.

In this way, this book is a collection of the papers by a sole author rather than a monograph. It consists of the newly revised and edited versions of the author's papers formerly published in Japanese or English, which are cited in footnotes throughout this book. (One paper which is the base of Chapter 6 was awarded "Outstanding Paper Award" by the Japan History of Science Society in 2018.) Therefore, after reading the introduction, the reader is free to read any chapter in any order.

The research that has crystalized in this book was supported by the Japan Society for the Promotion of Science, which provided the author its Grant-in-Aid for

Scientific Research seven times for the subjects related to this book, by the Nomura Foundation for Social Science, the Fukutake Science and Culture Foundation, the Mitsubishi Foundation, and by the Japan Securities Scholarship Foundation.

The author would also like to express his sincere thanks for the scientific support and encouragement from the late professor Kunioki Katô, the late professor Toshio Yamazaki, Prof. Masakatsu Yamazaki (Tokyo Institute of Technology; Tokyo Tech., Emeritus), Prof. Takeshi Tomita (Seikei University, Emeritus), Prof. Tsuyoshi Fujioka (Osaka University of Economics and Law, Extraordinary), Prof. Zenji Asaoka (Tôhoku University), Prof. Kôji Kanayama (Kyûshû University), Prof. Hirofumi Saitô (Tokyo Tech), Prof. Yakup Bektas (Tokyo Tech), Prof. Hideki Kan (Kyoto University of Foreign Studies), Prof. Tetsurô Katô (Hitotsubashi University, Emeritus), Prof. Yûji Wakao (Nagoya University, Emeritus), Prof. Ei-ichi Kido (Osaka University), Prof. Chieko Kojima (Nihon University), Prof. Yasushi Kakihara (Tokyo University of Marine Science and Technology), Dr. Hiroko Takahashi (Nagoya University), Prof. Toshihiro Higuchi (Georgetown University), Dr. Maika Nakao (Nagasaki University), Prof. Shunsuke Izuta (Dokkyô University), Prof. Takehiko Hashimoto (University of Tokyo), Prof. Shintarô Yoshimura (Hiroshima University), Prof. Emiko Nakasaka (Chûô University), Ms. Mayumi Furutani (Hiroshima University, Secretary), Prof. Vladimir Pavlovich Vizgin (Institute for the History of Natural Science and Technology, IIET), Dr. Yurii Mikahilovich Baturin (IIET), Prof. Eduard Izrailevich Kolchinskii (IIET, St. Petersburg), Dr. Tat'yana Ivanovna Yuspova (IIET, St. Petersburg), Ms. Larisa Petrovna Belozerova (IIET), Dr. Oksana Danirovna Simonenko (IIET), Dr. Konstantin Aleksandrovich Tomilin (IIET), Dr. Kirill Olegovich Rossiyanov (IIET), Dr. Vitalii Yur'evich Afiani (Archive of the Russian Academy of Sciences, ARAN), Dr. Nadezhda Mikhailovna Osipova (ARAN), Ms. Irina Georgievna Tarakanova (APAN), Ms. Natal'ya Sergeevna Prokhorenko (St. Petersburg Branch, ARAN), Dr. Stella Vladimirovna Pisareva (Museum of the History of Kazan' Federal University), Ms. Tamara Nikolaevna Martynova (Scientific Archive of the Siberian Branch of Russian Academy of Sciences), Ms. Natal'ya Pavlovna Sklyarova (Scientific Archive of the Ural Branch of Russian Academy of Sciences), Ms. Evgenia Grigorievna Sorokina (Central Municipal Archive of Moscow), Ms. Elena Aleksabdrovna Tyurina (Russian State Archive of Economy), Dr. Wolfgang Matz (Helmholtz Zentrum, Dresden-Rossendolf), Prof. Slava Gerovitch (Massachusetts Institute of Technology), Prof. Alexei Kojevnikov (British Columbia University), Prof. Nathan Brooks (New Mexico State University), Mr. Shengbin Tan (Taylor & Francis Asia Pacific), Mr. Simon Bates (Taylor & Francis Asia Pacific), Ms. Jennifer Bonnar (Apex CoVantage), and other colleagues, fellow historians, and friends.

Dear readers, please allow the author to dedicate this book to the memory of Masanori Kaji, an outstanding historian of chemistry, an authority on the study of Dmitrii Ivanovich Mendeleev and Vladimir Vernadskii in Japan, a professor at Tokyo Institute of Technology, and one of the author's closest friends, who passed away unexpectedly on July 18, 2016, at the age of only 60, due to pancreatic cancer.

May 2018, in Hiroshima,
Ichikawa, Hiroshi

Introduction

The research agenda and methods of the history of Soviet science and engineering in the early days of the Cold War

In accordance with the intensification of the Cold War, many scientific talents were mobilized in both the United States and the Soviet Union to several military-related research and developments such as nuclear weapons, rocketry developments, computer science, and others. Amid the intense, hostile ideological campaigns, several scientists were condemned, harassed, and purged not only in the Soviet Union, but also in the United States.[1] Eventually, mankind reached the dawn of the Atomic Age, with some nuclear power plants beginning to operate in the 1950s. In the Soviet Union, the people somewhat enjoyed the "thaw" after the death of the dictator, Iosif Stalin (1878–1953), and Soviet science showed high activity despite some anxiety about a possible reversal to the Stalin era.

Why were such a great number of scientific talents mobilized for various Cold War research and development projects? How can we explain that reason in the light of the newest achievements of the historical study, in which the simple totalitarian model has already lost credence? How can we characterize the basic features of the Soviet Cold War research and development projects in their realities? What were the real consequences of the rushed projects? How did Soviet scientists approach their nuclear age?

In resolving these questions, the remark of Friedrich Engels (1820–1895) at the funeral of Karl Marx (1818–1883) is suggestive: He identifies Marx's major contribution as his discovery of "the simple fact, hitherto concealed by an overgrowth of ideology, that mankind must first of all eat, drink, have shelter and clothing, before it can pursue politics, science, art, religion, etc."[2] We cannot fail to analyze material conditions of the people involved in the matters before analyzing the ideological aspects of the matters. Besides, it is suggestive that in the "Introduction" to a voluminous collective work on the Cold War science history, Naomi Oreskes questions

> how the Cold War shaped and altered the trajectories of science and existing technologies, how it created new sciences and technologies, how it affected the relationships between scientists and engineers and their patrons, and how scientists and engineers managed, negotiated, and adjusted those relationships – with varying degrees of success – as they attempted to achieve their own goals in relation to state patrons, fellow scientists (friendly or competitive), and personal aspirations.[3]

Also, for the United States' experience, having reexamined the history of computer development in the United States as one of the Cold War scientific efforts, Paul Cerruzi shed new light on the complicated interactions among various interest groups, objecting to the myth of "entire unity" and "foresightedness" of the military.[4] This book shares its viewpoints with these arguments.

The term "Soviet science" sounds somewhat odd. Science should not have any nationality, although some patterns of its development can be talked about with the national characteristics.[5] However, the term "Soviet science" is often used for the unique development of scientific studies in a relatively isolated environment, including the political, socio-cultural context within which such a model of scientific development was achieved. For a couple of decades and more, we have seen major changes in the historiography of Soviet science. Since the publication of the shocking book of Zhores Medvedev (1925–), *The Rise and Fall of T. D. Lysenko* (Columbia University Press, 1969), Soviet science and the scientific community have been depicted as passive victims of the Party-state's control. Such views, however, are currently being reexamined. The relationship between scientists and the political authorities was not dichotomous in its nature, but it involved a complicated interaction of various factors. One general characteristic of recent studies on the history of the former Soviet Union is the tendency to give increasing support to the understanding of its society based on the concept of "centralized pluralism," which Alec Nove (1915–1994), an economist, brought forward in a pioneering way. Already, in 1977, he recognized that "centralized decision making in a large modern industrially developed economy sets up an unmanageably large number of micro-economic interrelationships," especially because of "the ministers' insufficient resources and authorities owing to bureaucratic overlapping of organizations"[6] in the Soviet Union. Nove pointed out that "ministerial empires, and upward pressures originating with them (the inescapable delegation, devolution or decentralization – *Ichikawa*), are facts of economic and political life"[7] of the Soviet Union. Furthermore, also in the history of science, the traditional totalitarian model has been replaced by a more pluralistic viewpoint. Thanks to the study of the formerly classified documents in various archives, the totalitarian model that had been applied to the understanding of Soviet society for a long time lost its popularity rapidly and was replaced by a more pluralistic view. Seen from the outside, the path that Soviet scientists followed was full of "bizarre events." Nikolai Krementsov, however, mentions that "these events grew naturally out of the Stalinist system of science, a system whose contours we have understood in terms of the very same processes and modes of analysis – institutional structures, interactions among competing groups and individuals, and professional cultures – that Western historians have used to understand their own science."[8] He, therefore, focuses on institutional structures, interactions among competing groups and individuals, and professional cultures as the background for those "bizarre events." We can mention some remarkable historians of science belonging to the new trend – Nikolai Krementsov, the author of *Stalinist Science* (Princeton University Press, 1997), Slava Gerovitch, the author of *From Newspeak to Cyberspeak: A History of Soviet Cybernetics* (The MIT Press,

2002), and Alexei Kojevnikov, the author of *Stalin's Great Science: The Times and Adventures of Soviet Physicists* (Imperial College Press, 2004). They are all remarkable works, pioneering the new way of the history of Soviet science. Most of them, however, limit their view to certain fields of science and are apt to focus on some prominent scientists (their biographical facts), according to the traditional methodology of the history of science. In addition to this, they seem to be apt to focus on the late Stalinist era with an exception of Slava Gerovitch and, to some extent, Alexei Kojevnikov, because the greatest scientific efforts were made, many "bizarre events" took place, and Soviet science flourished in that era.

The performance of Soviet science in the Cold War, especially in 1950s, is a consequence of complicated interactions among various factors and actors, such as the indigenous contradictions in the historical development of science in the Soviet Union, the conflicts among interest groups, their relationship to the political and/or military leadership, the ambitions, the expectations, the careerism, and/or the morals of politicians, bureaucrats, scientists, engineers, and others involved in the Cold War scientific and technological projects. To resolve "the simultaneous equations" composed of such "variables," we must also analyze the behaviors of a wider range of "the actors," such as many ordinary, less well-known scientists, engineers, bureaucrats, military officers, and others from the viewpoint of their material lives, without exaggerating the ideological aspects. At the same time, however, we must take into consideration that all these variables were related to the Cold War and all the actors were, to some extent, inspired by the Cold War. We can converge analysis of all the above-mentioned factors and actors on the Cold War phenomena. This book is such an endeavor.[9]

In following pages, we shall take a glance at the problematics concerning the topics we take up in this book.

What drove the scientists to the Cold War?

The younger generation of scientists, like Arkadii Brish (1917–2016), who grew up in the Soviet era and participated in the early Soviet nuclear weapon development naively believed in the justice of the defense of their own homeland and/or the doctrine of the equilibrium of nuclear military power.[10] However, according to a recent study, Lev Landau (1908–1968. A Nobel Laureate in Physics in 1962) was a convinced, anti-Stalinist activist.[11] Outstanding physicist Evgenii Zavoiskii (1907–1976) was worried about the moral and ethical aspects of the project in which he took part and looked for a chance to escape from it.[12] Why did the political power succeed in involving many physicists, including Landau, Zavoiskii, and other politically incredible persons? Or, more exactly, what drove them to the Cold War scientific projects? No one can answer this question briefly. However, here we should give enough thought to the material and ideological environment that surrounded Soviet physicists in those days.

Oreskes calls the readers' attention to the problem, the so-called "Cold War climate," including "a significant component of hysteria," citing Patrick Blackett (1897–1974. A Nobel Prize laureate in Physics in 1948).[13] Independently from Ethan Pollock,[14] the author has taken notice of the consequences of an ideologically charged campaign under the guise of a series of academic "discussions" initiated in various fields of science in the postwar Soviet Union, as one of the important occasions for cultivation of "the Cold War climate" among Soviet physicists.[15] Soon after the "discussion" began first in the field of philosophy in 1947, and also in the field of physics, Vladimir Kessenikh (1903–1970), Vasilii Nozdryov (1913–1995), and several other professors and teachers in the Faculty of Physics (*Fizicheskii fakul'tet*; Fizfak) of the Moscow State University (MSU), who later called themselves "patriotic and materialistic physicists," launched an attack on such leading physicists as Grigorii Landsberg (1890–1957), Pyotr Kapitsa (1894–1984. A Nobel Prize laureate in 1978), and Igor Tamm (1895–1971. A Nobel Prize laureate in 1958), condemning them for their "idealism" and "cosmopolitanism or worship of foreign science."[16] Immediately after the notorious "August Session" of the All-Union Academy of Agricultural Sciences named after V.I. Lenin (*Vsesoyuznaya Akademiya Sel'skokhozyaistvennykh nauk imeni V.I. Lenina*; VASKhNIL), in a jointly signed letter dated December 3, 1948, Sergei Kaftanov (1910–1978), Minister for Higher Education, and Sergei Vavilov (1890–1953), President of the USSR Academy of Sciences, called on physicists at research institutions and education facilities across the Soviet Union to attend a "conference" to discuss the struggle against "idealism in physics" and "cosmopolitanism or worship of foreign science."[17] Although this conference was never held, the ideological dispute between the two groups of Soviet physicists, mainly in the preparatory stages of the conference, was severe and intense. This issue has been extensively analyzed by Ethan Pollock and other outstanding historians of science, such as Vladimir Vizgin[18] and Anatolii Sonin.[19] Putting stress on Stalin's ambition to be the "coryphaeus of science" as the major factor behind such academic "discussions," Pollock thoroughly clarifies the tremendous complexity, ambiguity, and multiplicity of the academic discussions. He is correct when he says the "outcome from one discussion did not translate into clear lessons for other disciplines,"[20] and "there is no extant evidence that Stalin coordinated the scientific discussions with any specific diplomatic or military initiatives of early postwar period."[21]

Given, however, that most participants in that dispute were staff members of Fizfak despite the difference of their status, full-time or part-time, it can aptly be described as "a tempest in a tea cup." Many of the accusers, Fizfak physicists, conducted the philosophical disputes with reckless valor despite their own lack of philosophical knowledge. It is, however, difficult to say whether historical studies of Soviet science have sufficiently analyzed the "tea cup" itself, i.e., the circumstances surrounding the community of physicists at Fizfak. Prior to this dispute, they had gone through a formidable wartime evacuation during World War II. Even after returning to Moscow toward the end of the war, the recovery of their research facilities from the war damages and their adaptation to the new circumstances took time and effort.[22]

A general view of the wartime efforts made by Soviet scientists was given by the comprehensive work by Boris Levshin (1926–2012).[23] When examining the impact of the war on the Soviet scientific community, the main assumption of Soviet historiography on the wartime activities of Soviet scientists has been the claim that Soviet scientists were fighting against the fascist invaders.[24] Despite many new analyses based on the documents newly declassified after the collapse of the Soviet Union, the writings on the history of Soviet science in the war period continue to follow a general trend of glorifying the feats of Soviet scientists. We can mention the elaborate work by E.I. Grakina as a remarkable example of such a trend.[25] A more credible line is taken by another group of historians, for instance, Nikolai Krementsov, who asserts that, during the war, scientists strengthened their autonomy and established a "symbiotic" relationship with political authorities.[26]

It seems that these new studies have considered only those scientists and organizations that made significant efforts in wartime research and development.[27] It should be noted, however, that outside of the Academy of Sciences, there were many scientists who could not make such major contributions in science. What is more important, it seems that these studies have left unanswered an important question: Why did leading scientists who had made such significant contributions to the war efforts have to be condemned immediately after the war?

Some aspects of Soviet Cold War research and developments

Sonja Schmid questions:

> How could a country that had within the past 50 years gone through the political turmoil of empire collapse, back-to-back revolutions, and civil war, a country whose population had suffered devastating famine during the forced collectivization of agriculture and horrendous loss of life during World War II, succeed in cutting-edge science and innovative technology?[28]

At the very beginning of its postwar period, the Soviet Union successfully developed a number of new military technologies – nuclear weapons, jet aircraft, nuclear submarines, advanced radars, missiles, etc. – and reequipped the defense industry on a large scale. Considering the serious damage suffered by the whole industry during the war, we must understand that such rearmament was to constitute a very hard drive.

On May 13, 1946, a special resolution of the Council of Ministers of the USSR charged the Ministry of Armament, which was headed by Dmitrii Ustinov (1908–1984), with the responsibility of developing long-range missiles. It should be noted that, in the prewar period, rocket engines were developed in the Soviet Union by other organizations – the Central Aero-hydrodynamic Institute (*Tsentral'nyi Aerogidrodinamicheskii Institute*; TsAGI) and the Reactive[29] Research Institute (*Reaktivnyi nauchno-issledovatel'skii institut*; RNII). Sergei Korolyov (1907–1966) and Valentin Glushko (1908–1984), who worked in these organizations, were later involved in the rocket development project under the Ministry of Armament. The Ministry of Armament, which began to direct the rocket development project immediately after the war, did not have any achievements before in this area.

In addition, the Ministry of Armament started developing a long-range missile by merely copying the German "Vau-2 (V-2)." Despite the great influence of the V-2 on the forthcoming missile age, Mark Harrison suggested that the V-2, loaded with a conventional warhead, was inaccurate, unstable, limited in scope, too expensive, and not effective for mass destruction of the cities of Allied Power nations during World War II. Then why did they make such a start? Harrison, cautiously assessing the various aspects of the issue, concluded that the Soviet military industry had to reproduce the V-2 first, because of the desperate backwardness of Soviet industry.[30] However, the leading Soviet engineers – Korolyov, Boris Chertok (1912–2011), and others – opposed the policy of mere copying. Asif Siddiqi, who attracted considerable attention by his detailed reexamination of and fresh approach to the overall history of Soviet rocket development and space exploration, seems to have paid relatively small attention to this question; he regards the importance of V-2 as self-evident, saying, "All roads, in other words, led to the German rockets, and from the German rockets sprang the future."[31]

General Lev Gaidukov (1911–1999), Chair of Technical Commission on Reactive Armaments affiliated to the Military Administration in Berlin, was asked by the Party's Central Committee to recommend organizations suitable for research on rocket technology soon after the establishment of his commission. He asked some leaders of the Soviet defense industry whether they were interested in rocket development: Aleksandr Shakhurin (1904–1975), Minister of Aviation Industry; Boris Vannikov (1897–1962), Minister of Ammunition, whose ministry was soon to be abolished; and Dmitrii Ustinov. Only Ustinov accepted Gaidukov's proposal after consulting with his deputy, Vasilii Ryabikov (1907–1974).

The author of D.F. Ustinov's biography considers this decision to be the result of his foresight.[32] In fact, Ustinov, who attracted a number of state leaders, including Lavrentii Beria (1899–1953) and Georgii Malen'kov (1902–1988), actively urged Stalin to entrust his ministry with the responsibility of missile construction.[33] Here, we must go behind "the myth of foresightedness."

As for the development of jet planes in the Soviet Union, it was, no doubt, a process of a huge scale of "plunder of technology" from a country defeated in the war orchestrated by a victorious country on an unprecedented scale.

The development of military technology in the former Soviet Union immediately after the war employed many German engineers the Soviet Union acquired "for free." Therefore, the process of acquisition and utilization of German technology is one of the essential elements in the development of military technology in the former Soviet Union.

It is known that, since the beginning of the occupation of Germany, the Soviet Union government established research and development facilities within German territory and mobilized many local specialists to learn German jet plane technology. For example, based on the statements of German specialists mobilized for research and development from a very early period, Ulrich Albrecht[34] considers this a process undertaken by the Soviet Union to copy German technology. Moreover, Russian historians Sergei Kuvshinov and Dmitrii Sobolev[35] traced the activities of German specialists such as Brunolf Baade (1904–1969) before and after they were taken to the Soviet Union and confirmed their immense contribution to the jet plane technology of the Soviet Union. Recently, in his reflective overview on history,[36] Nikolai Stroev provided an accurate description of the use of "war trophies" acquired from Germany, and, in a study related to the history of development of the military industry of the Soviet Union, Nikolai Simonov[37] described the research and development process of jet plane technology from an organizational aspect, using numerous documents.

However, despite the attempt at a large-scale "technology transfer" from Germany and the fact that the upper echelons of the government wanted a copy of Germany's airplanes, contrary to the case of rocket development, the initial jet planes produced by the Soviet Union were not simple copies of German airplanes. What were the reasons behind this outcome? To what extent did the Soviet Union succeed at learning German technology, and how limited was it? The details of jet plane development were out of the scope of the aforementioned studies, which, therefore, cannot answer these questions completely.

Considering these questions, we will examine the process and problems of the independent research and development conducted in the former Soviet Union until the end of the war, the development of research and development facilities in Germany, and their research contents and limitations. We will also analyze the process of Soviet withdrawal from German territory (which corresponds to the transfer to the Soviet Union's territory).

During the Cold War, the United States devoted enormous financial and human resources to the production of high-speed computers. The United States created sophisticated computer technology, on which our modern computer technology today depends.[38]

Compared with that of the United States, the Soviet initiative to produce an alternative high-speed computer achieved little. As historians have rightly pointed out, the Soviet Union was highly successful with the production of rockets, jet planes, and nuclear weapons, but, in comparison, its attempt to design a high-speed computer was the least successful undertaking.[39]

Some believe that an anti-cybernetics campaign, which emerged in the early days of Soviet computing, "significantly delayed the development of Soviet computers."[40] Contrary to this belief, Gerovitch shows that "Soviet computing was shaped by the tension between the practical goal of building major components of modern sophisticated weapons and the ideological urge to combat alien influences."[41] He argues that while the "soldiers of the ideological front" (the philosophers and the Party ideologues) were dismissing cybernetics as a "modish pseudo-science," "the actual soldiers in uniform" (the highly ranked military officers) took Western military research on computing and control very seriously.[42]

Despite the official date of the completion of the "first" Soviet digital computer, BESM (*Bystrodeistvuyushchaya Elektronnaya Schetnaya Mashina*; a high-speed electronic calculating machine), in 1952,[43] the machine in question was actually not completed until 1955. The experimental work on the first BESM model continued throughout the following year.[44] It passed the State Efficiency Test only in 1955. Another early Soviet digital computer, Strela-1 (or Arrow) passed the same test a year earlier (1954), and it was put into practical use in no time.[45] The designers of Strela-1 were all awarded State Prizes.[46] Its chief designer, Yurii Bazilevskii (1912–1983), was, in particular, heralded as a Hero of Socialist Labour.[47] However, two years later, Strela was abruptly abandoned with only seven machines completed.[48] How can we understand that entangled process?

Sergei Prokhorov gives us a well-arranged overview of the history of Soviet computer development and its institutional framework.[49] In their elaborate study, George D. Crowe and Seymour E. Goodman reveal the details of the BESM computer development and the rivalry between the two institutions involved in Soviet computer development.[50] In his remarkable monograph, Slava Gerovitch offers an interpretation of that rivalry.[51] Citing the instance of an objection from the Ministry of Machine-building and Instrument-building (*Ministerstvo Mashinostroeniya i Priborostroeniya*; MMiP) against the opening of BESM to a mission of Indian scientists and focusing on the attitude of the governmental agency (in this case, MMiP) toward disclosing the information, he concludes, "This case suggests, furthermore, that the policy of secrecy pursued by the Soviet state was not solely the product of Soviet isolationist ideology, but could sometimes be induced by interagency rivalry and used as a weapon of bureaucratic competition."[52] The problem of rivalry is, of course, of importance. I believe, however, that what is the most important here is the way through which the necessity of developing high-speed, electronic, digital computers in the Soviet Union came into being. So far, however, there is insufficient knowledge of Strela and the institution that built it. Therefore, the answer to the above-mentioned question remains unclear.

This book (Chapter 5) provides the interests and the behaviors of the engineers and the scientists involved in the initial developments of the Soviet high-speed, digital computers. What is found is what can be described as a conflict of interests between the mechanical engineers and the mathematicians. Whereas the former were associated with the Ministry producing the measuring and the controlling instruments for the artillery, the latter were identified with the nuclear developments. There is little evidence of the influence of the ideological campaign. The Soviet institutions engaged in the development of high-speed electronic digital computers were the Institute of Precision Mechanics and Computing Technique (*Institut Tochnoi Mekahniki i Vychislichel'noi Tekhniki*; ITMiVT) of the Russian Academy of Sciences, which later developed BESM computers, and Special Design Bureau No. 245 (*Spetsial'noe Konstruktorskoe Byuro No. 245*; SKB-245) of MMiP, which later developed the Strela machines.[53]

First, there is a need for reexamination of the initial attitude of Soviet mechanical engineers toward computing technology and their main orientation with the activities of ITMiVT in its early days and of SKB-245, whose main clients were the artillerist part of the army and other military organizations mainly concerned with trajectory and aeronautical dynamism. At the same time, we must analyze the internal discord between the mechanical engineers and the mathematicians within the Academy of Sciences, especially in its Department of Technological Sciences, and the consequences of that conflict. Finally, we examine the counteraction from the side of SKB-245, which once lost its advantage over the newly reorganized ITMiVT.

"Hope" and "anxiety": Soviet science at the dawn of the nuclear age

Paul Josephson, a well-known historian who has long been engaged in the history of Soviet science, stated in his well-received book, *Red Atom* (2005), that the Obninsk Atomic Power Station, the "site of the first reactor to produce electricity for a national grid in 1954," beat "any United States efforts to commercialize nuclear power by four years."[54] He went on to say,

> The Obninsk power station was a propaganda coup for the Soviet Union. It demonstrates the peaceful intensions of the nation on the heels of U.S. President Dwight Eisenhower's address to the United Nations calling for shared nuclear know-how in medicine, agriculture, transportation, and power generation for the benefit of all humankind.[55]

The success of the Obninsk Atomic Power Station caused further difficulties for the U.S. government, which "had been perplexed at the not so successful realization of the 'Atoms for Peace' address by President Eisenhower."[56]

On July 1–5, 1955, the Academy of Sciences hosted a large-scale scientific conference in Moscow called the Session of the Academy of Sciences of the USSR on the Peaceful Use of Atomic Energy (Hereafter, the Session).[57] With many foreign scientists invited, it was an international stage to debut the Soviet-made nuclear reactor built for peaceful purposes – the first such reactor in world history – and to demonstrate the achievements of Soviet science. The Session was immediately followed by the well-known First United Nations International Conference on the Peaceful Use of Atomic Energy (Hereafter, "the Conference") held in Geneva on August 8–20 of the same year, in which the Soviet scientific achievements in this area rivaled those of the United States.

Nevertheless, historians have rarely discussed the Session prior to the Conference, the efforts that the Soviet Union put into the paired events, and the self-awareness that Soviet scientists acquired through these events. Some exceptions include David Holloway's and John Krige's comments concerning both events. Holloway referred to them in his well-accepted monograph, *Stalin and the Bomb* (1994). He pointed out that the conduct of the Session and the positive participation of Soviet scientists in the Geneva Conference were backed by their strong desire to reestablish international liaisons with Western scientific communities and internationalize Soviet science after the long interval of more than two decades since the repressions in the 1930s. He mentions the initiative to prepare for these paired events taken by Igor Kurchatov (1903–1960), the leader of Soviet nuclear development as a whole, who believed "that Soviet nuclear physics was suffering from unnecessary secrecy and from the lack of contact with Western physicists."[58] Under the assumption that the internationalization of scientific activities is often based on some nationalistic intentions, John Krige concludes that "The Atoms for Peace Conference in Geneva in 1955 was such a panopticon,"[59] i.e., an arena for scientific intelligence gathering for both sides of the Cold War.

However, neither Holloway nor Krige clarified the consequences of the Session and the Conference in the context of Soviet scientists because of the shortage of research resources regarding this topic. Krige says, "We do not know how Soviet scientists and their administration used the information they got from the British and the Americans in Geneva to strengthen and reorient their national nuclear programs."[60]

In this book (Chapter 6) the author tries to shed new light on these paired events, the Session and the Conference, after reexamining the first Soviet reactor for commercial use at the Obninsk Atomic Power Station.[61]

The advent of the Atomic Age, mainly its peaceful use, means, at the same time, the beginning of the anxiety regarding the radiation effects on living bodies.

In June 1957, a Japanese radiologist and peace movement activist, Yasushi Nishiwaki (1917–2011), visited Moscow and gave a lecture warning of increases in radioactive fallout from nuclear tests. Shortly before his talk, he also wrote an open letter to the Soviet political leader of that time, Nikita Khrushchev (1894–1971), with the request for his statement announcing the Soviet's readiness to accept the ban on nuclear testing if the Western countries stopped theirs.[62] Nishiwaki's visit to the Soviet Union coincided with the moment when Soviet scientists stood at the crossroads in understanding radiation effects on the living body. Their perspective was changing from that of a mere follower of Western studies to that of a radical critic. In 1959, following the serious debates on radiation effects at the United Nations Scientific Committee on the Effects of Atomic Radiation (UNSCEAR), some Soviet scientists assembled their critiques as a book. Entitled *Sovetskie Uchyonye ob Opasnosti Ispytanii Yadernogo Oruzhiya* (*Soviet Scientists on the Danger of Nuclear Weapon Tests*),[63] this coedited volume covered a wide range of topics in great detail. Among them were: the mechanism of internal exposure, skepticism about the "averaging method" adopted in Western studies, difference in effects among various countries because of diversity in diet,[64] the formation of radioactive carbon in the air and its effects, and the concentrations of radioactive particles and their long-term effects.

The timing of their criticism was, however, crucial for understanding the context of this drastic shift in Soviet scientific opinions on radiation hazards. Only in the summer of 1958 did the Soviet scientists begin to present their critical views on the Western risk assessments regarding radiation protection. Their work criticized the reports and recommendations made by the International Committee on Radiation Protection (ICRP). This was despite the fact that ICRP had been in operation since 1950, and UNSCEAR since 1954. Toshihiro Higuchi has explained the reason for this delay, arguing: "Another reason why the Soviet Union fell behind the Western countries in taking part in the international risk assessment activities was that a reconstruction of knowledge was then going on in the Soviet Union starting with the radiation effect problem."[65] What is most striking in the history of Soviet science is, no doubt, the dominance of a pseudo-science, Lysenkoism. Even in those days, Soviet genetics was dominated by the Lysenkoite doctrine that denies gene theory. However, the leading Soviet atomic scientists made every effort to establish a research center for biological and genetic studies competing with the Lysenkoites. Independently from Higuchi, the author has also made a similar argument that the subject of radiation effects on human heredity had to be linked to the "normalization" of the biological sciences in the Soviet Union at the time when Lysenkoism was still rampant. Moreover, he has pointed out that the normalization of biological sciences was one of the targets of the de-Stalinization efforts launched by the leading Soviet scientists immediately after the death of the dictator Stalin.[66]

Such a view, however, might lose sight of some crucial points. The assumption behind this view – the convergence of the interests between political power, seeking to mobilize scientific knowledge against the U.S. and U.K. initiatives in the field of radiation studies, and scientists attempting to normalize biological sciences – may be an oversimplification, considering the current tide in Soviet studies. Indeed, despite intense opposition among scientists, and their subsequent weakening, the Lysenkoites managed to maintain their dominance in Soviet biological sciences until the downfall of Khrushchev in 1964.[67] An even more important point is that "cooperation" between the Kremlin and scientists might have been a double-edged sword because of the difference in timescale between the transient nature of governmental policies and the long-term commitment required for scientific research. Indeed, Andrei Sakharov (1921–1989. A Nobel Peace Prize laureate in 1975), who was mobilized to level criticisms against the Western methods for radiation risk assessment, grew very skeptical of the wisdom of the nuclear weapons development in his own country.[68]

Therefore, we must look for another context in which research on radiation effects emerged as a major topic in Soviet science. This book (Chapter 7) aims to explain the indigenous needs for such research in the Soviet Union, as well as its early development by shedding light on radiation casualties at a nuclear development center, Chelyabinsk-40 (later, -65) and hazard studies conducted there in its early years. In Chapter 7, at first, we shall take a close look at the serious situations in Chelyabinsk-40 to see how research on radiation effects was needed to solve practical problems from the very beginning of the Soviet nuclear project. Second, we shall examine the process and features of the indigenous and original research on radiation effects in the Soviet Union at the initial stage of its development. Finally, we shall focus on the course of events unfolding within the Soviet Academy of Science, which then had an overwhelming influence and a very high authority among the Soviet scientists, to examine the reality of the Lysenkoites' hegemony in the circle of scientific leaders. This inquiry will be limited to the period from the establishment of the first nuclear facilities to the summer of 1958, when the heated debate began in UNSCEAR.

The rise of radiation effect studies in the Soviet Union also bore fruits for the scientists. The downfall of Trofim Lysenko (1898–1976) and the normalization of biological sciences was, without any doubt, a result of the de-Stalinization conducted by the then leading scientists, which began immediately after the unexpected death of the dictator, while they, the scientists, were engaged in some Cold War scientific projects. Stalin, perhaps inspired by Lamarckian doctrine, supported Lysenko with some reservations against Lysenko's exaggeratedly ideological remarks.[69] Because of their victory at the August Session of VASKhNIL, the Lysenkoites had maintained their dominance in biological sciences until 1965, one year after the downfall of their patron Khrushchev. Khrushchev, as First Secretary of the Central Committee of the Communist Party of USSR and Premier of the USSR, faced with the crisis in Soviet agriculture, came to expect much from Lysenkoite agronomy, which had provided a variety of "cheap methods" to raise productivity in agriculture. He made his support for Lysenko clear in 1957 and condemned Nikolai Dubinin (1907–1998), the main figure in the opposition against Lysenkoites, at a general assembly of the Party's Central Committee on July 29, 1959.[70]

Nevertheless, the Institute for Cytology and Genetics of the Siberian Branch of the USSR Academy of Sciences with Dubinin as the head was established in 1957, when Lysenko obtained support from the top political authorities again. In addition, that institute had a large body of members, even at the initial stage, and continued to expand in scale despite interference from Lysenkoites. Eventually, as Paul Josephson and Sergei Shalimov described in detail, that institute played the main role for rescue, rehabilitation, and development of genetics in the Soviet Union.[71]

Why was the coincidence of the Lysenkoites' hegemony and the establishment of "Dubinin's Institute" possible? Valerii Soifer (1936–) refers to "some complexity of Khrushchev Era" in search for the reason why such a coincidence became possible.[72] Although the control of the political power over science was still accepted in those days, Soviet scientists believed in the irreversibility to Stalinist era and took diverse directions.

Slava Gerovich, having attracted worldwide attention by his fresh approach to the flourishing of cybernetics in the Soviet Union in the 1960s as a sort of social reform movement led by scientists, takes notice of the role of a mathematician Aleksei Lyapunov (1911–1973), who invited biologists, including Dubinin, to a private circle for the sake of his two daughters with ambition to specialize in biology. Thus, eventually, Lyapunov became convinced of the indispensability of the normalization of Soviet biology, i.e., breaking down the Lysenkoites' hegemony, as the most important problem for Soviet science and took action for this purpose. One of the fruits of his activities was the so-called "Letter of Three Hundreds," which appealed with the signatures of 297 scientists to the Presidium of Party's Central Committee for the normalization of biological sciences.[73] Gerovitch believes that such a "social movement" attended by a wide range of scientists for the normalization of biology as an undercurrent prepared the positive conversion of the situation in biology, such as the establishment of a new research institute with Dubinin as its Director.

However, we must take into consideration that the Institute for Cytology and Genetics of the Siberian Branch of Soviet Academy of Sciences was established by the Academy of Sciences as an official organization with a high level of autonomy. The Institute's administration and activities were managed by the Academy's Siberian Branch relatively independent from the Academy's headquarters in Moscow. We must take into consideration this institutional aspect. In Chapter 7 of this book the author sheds light on the enigma of the coincidence of the restored Lysenkoites' hegemony and the establishment of the "Dubinin's Institute," by examining the minutes and stenographic notes of the Academy's Presidium in Moscow and the records and documents of the Academy's Siberian Branch in Novosibirsk.

We must also take into consideration that genetics, or more widely biology, which were then facing difficulties with the rise of molecular genetics and biology, could not help becoming a crucial topic in the arguments in the meetings of the Academy's Presidium, as the research of radiation effects on living bodies became increasingly important in the second half of the 1950s in accordance with the intensification of the nuclear arms race between the United States and the Soviet Union. First, we examine the importance of radiation studies at the time. We then examine the arguments of the prominent scientists within the Academy's organizations and the interference from the Lysenkoites and their consequences.

In 1959, the findings on the radiation effects by the Soviet scientists were crystalized in a single work, titled *Sovetskie Uchyonye ob Opasnosti Ispytanii Yadernogo Oruzhiya* (*Soviet Scientists on the Danger of Nuclear Weapon Tests*, cited in note 63). However, as discussed later, although these scientists had received data from Japan on the effects of radiation exposure accompanying the nuclear explosions in Hiroshima, Nagasaki, and Bikini Atoll, there is no sign that the scientists took full advantage of these data.[74]

In the late 1950s, there were many Soviet public-health-related publications that discussed radiation sickness, one of which was the pamphlet issued in 1957, *Luchevaya Bolezn': Posledstviya Porazheniya Organizma pri Atomnom Vzryve* (*Radiation Sickness: The Consequences of Biological Organ Damage Caused by Nuclear Explosion*) by Mikhail Pobedinskii (1900–1970).[75] Intended as a guide to prepare for the break-out of nuclear war, a looming possibility at the time, the pamphlet contained explanatory content for medical and public health specialists regarding the harmful effects of radiation and how to prevent them. It also featured photographs of and data on the *hibakusha* of Hiroshima and Nagasaki.

The eminent radiologist Anna Kozlova (1906–1980) visited Japan before the normalization of Soviet–Japanese diplomatic relations and brought back with her to the Soviet Union large amounts of data on radiation exposure from the Hiroshima, Nagasaki, and Bikini Atoll bombings. Here, the author discusses some little-known facts about Dr. Kozlova, including her biography, her stance on the problem of radiation, and how she regarded the data that she brought back with her, with a view to elucidating how the Japanese data on radiation exposure, including the direct radioactive effects of nuclear bomb explosions and of the effects of subsequent radiation exposure in Hiroshima, Nagasaki, and Bikini Atoll, ended up being used in the Soviet Union amid the above-mentioned bifurcation of the Soviet scientific community.

Notes

1 For example, see, Jessica Wang, *American Science in an Age of Anxiety: Scientists, Anti-Communism, and the Cold War*. University of North Carolina Press, 1998.
2 Friedrich Engels, "Karl Max's Funeral." www.marxists.org/archive/marx/works/1883/death/dersoz1.htm.
3 Naomi Oreskes, "Introduction." in Naomi Oreskes and John Krige (Ed.), *Science and Technology in the Global Cold War*. Cambridge, MA, London, The MIT Press, 2014, 3.
4 See, Paul Cerruzi, *A History of Modern Computing*. Cambridge, MA, London, The MIT Press, 1998.
5 Under the impact of World War I, the state sponsorship and governmental organization of scientific studies progressed on a large scale in various industrial countries. Before that "nationality of science" could be clearly characterized as "the amateur tradition" for Great Britain, "the entrepreneurship for invention and patents" for the United States, "the early sponsorship of the state" for German states, and so on [see, Tetsu Hiroshige, *Kagaku no Shakai-shi – A Social History of Science*. Tokyo, Chûô-KôRon-sha, 1973 (in Japanese)].
6 A. Nove, *The Soviet Economic System*. London, George Allen & Unwin, 1977, 56.
7 Ibid., 64.
8 N. Krementsov, *Stalinist Science*. Princeton and Oxford, Princeton University Press, 1997, 287.
9 This book, however, omits the topic of the early Soviet nuclear weapon development, the main event of the Soviet Cold War research and development efforts, because the restriction of access to the primary research source and other factors made it difficult for me to present any new and original arguments well grounded in reliable sources other than those given by the previous research, beginning with David Holloway's remarkable, comprehensive work, (David Holloway, *Stalin and the Bomb: The Soviet Union and Atomic Energy, 1939–1956*. Yale University Press, 1994), although I introduced to Japanese readers the outline of the history of early Soviet nuclear weapon development mainly based on the authoritative, comprehensive history published in Russia [Pod otvet. red. V.N. Mikhailova (Ed.), *Sozdanie Pervoi Sovietskoi Yadernoi Bomby* (*Making of the First Soviet Nuclear Bomb*). Moskva, Energoatomizdat, 1995g.] in my book published previously in Japanese [Hiroshi Ichikawa, *Reisen to Kagaku-Gijutsu: Sorenpô 1945–1955* (*The Cold War, Science and Technology: The Soviet Union, 1945–1955*). Kyoto, Minerva Shobô, 2007: see Chapter 1 in this book (15–97)]. For the same reason, the topics of some important projects, such as nuclear submarines and the radar network, are also omitted.
10 V.P. Vizgin, "Nravstvennyi vybor i otvetstvennost' uchyonogo-yadelshchika v istorii sovetskogo atomnogo proekta." *Voprosy Istorii Estestvoznaniya i Tekhniki*, No. 3, 1998, 110.
11 Gennady Gorelik, "The Top Secret Life of Lev Landau." *Scientific American*, August 1997, 72–77.
12 Vizgin, *op. cit.*, in note 10. 111.
13 Naomi Oreskes, "Science in the Origins of the Cold War." in Oreskes and Krige (Ed.), *op. cit.*, in note 1. 12: Also see, P.M.S. Blackett, *Military and Political Consequences of Atomic Energy*. London, Whittlesey House, 1949.

14 Ethan Pollock, *Stalin and the Soviet Science Wars*. Princeton and Oxford, Princeton University Press, 2006.
15 Ichikawa, *Reisen to Kagaku-Gijutsu . . . op. cit.*, in note 9. See the Chapter 2 in this book (99–153).
16 V.P. Vizgin, "Yadernyi shchit v 'Trinadtsatiletnei voine' fizikov s nevezhestvennoi kkritikoi sovremennykh fizicheskikh teorii." *Uspekhi Fizicheskikh Nauk*, Vol. 169, 1999, 1372.
17 Ibid., 1373.
18 Ibid., 1363–1389.
19 A.S. Sonin, *"Fizicheskii Idealizm": Istoria Odnoi Ideologicheskoi Kampanii*. Moscow, Fiziko-Matmaticheskaya Literatura, 1994.
20 Pollock, *op. cit.*, in note 16. 9.
21 Ibid., 215.
22 For example, Sonin, focusing on the personalities and thoughts of Soviet physicists and not giving so much attention to the material conditions surrounding them, considers the wartime merely as an armistice in ideological struggles (see, Sonin, *op. cit.*).
23 B.V. Levshin, *Sovetskaya Nauka v Gody Velikoi Otechestvennoi Voiny*. Moscow, Nauka, 1983.
24 Among Western historians, working on the theme of the history of Soviet science, Alexander Vucinich also believes that, despite the huge material and financial damage to Soviet science mainly due to the loss of research equipment and research bases and despite the huge damage to the lives and health of scientists, Soviet science in general increased its credibility in Soviet society. Vucinich himself was much influenced by the above-mentioned tendency in Russia, however (Alexander Vucinich, *Empire of Knowledge: The Academy of Sciences of the USSR 1917–1970*. Berkeley, Los Angels, London, University of California Press, 1984).
25 E.I. Grakina, *Uchyonye Rossii v Gody Velikoi Otechestvennoi Voiny*. Moscow, Institut Rossiiskoi istorii, 2000.
26 Nikolai Krementsov, *op. cit.*, in note 8. 193–226. As for the impact of the war, B.I. Kozlov makes an interesting point that "decentralization" of the control system, brought about by the disorder accompanying the German invasion, made the leading scientists of the Academy of Sciences grope for a solution in the dark; they began to follow the precedent of World War I (B.I. Kozlov, *Akademiya Nauk SSSR i Industrializatsiya Rossii: Ocherk Sotsial'noi Istorii 1925–1963*. Moscow, Academia, 2003, 142, 143).
27 Krementsov emphasizes two important changes in Soviet society that were brought about by the war: the loss of "overconfidence" in the Party and state bureaucrats, and the proportional rise of the importance of and the confidence in specialists in various fields of Soviet society. He also points to the emergence of the "gaps" in the ideological aspect of party life, associated with the adoption of many new members in the campaign to join the party during the war. Scientists strengthened their autonomy and established a "symbiotic" relationship with the political authorities, when several academic scientists were appointed to positions in the leaderships of some People's Commissariat or became officers of high rank. Krementsov does not give many examples to support this view, however (Krementsov, *op. cit.*, in note 8. 96–99).
28 Sonja D. Schmid, "Defending (Scientific) Direction: Soviet Nuclear Physics and Reactor Engineering During the Cold War." in Oreskes and Krige (Ed.), *op. cit.*, in note 3. 318.
29 In Russian language, the words, "a jet engine" and "a rocket engine" are lumped together under the words *reativnaya dvigatel'* (a reactive engine), because that type of engine makes use of reactive force by heat expansion of gas.

30 M. Harrison, "New Postwar Branches (1): Rocketry." in J. Barber and M. Harrison (Ed.), *The Soviet Defense-Industry Complex from Stalin to Khrushchev*. London, Macmillan Press, 2000, 126–130.
31 Asif Siddiqi, *Red Rocket Clare: Spaceflight and the Soviet Imagination, 1857–1957*. Cambridge, Cambridge University Press, 2010, 197.
32 See, Yu.S. Ustinov, *Narkom, Ministr, Marshal*. Moscow, Izd-vo 'Patriot', 2002, 93–96; Yu.S. Ustinov, *Trizhdy Geroi i Kavaler*. Moscow, Nauka, 2005, 115–185.
33 See, Ya.K. Golovanov, *Korolyov: Fakty i Mify*. Moscow, Nauka, 1994, 362–364. The author of this book supposes that Ustinov expected diffusion of an artillerist radar, SON, in the development of which his People's Commissariat engaged. Beriya, who consistently supported Ustinov, subsequently planned to make a brilliant debut of his son, Sergei (Sergo), as an outstanding engineer in radio engineering when creating an air defense system (see, A. Dokuchaev, "Gornaya Taina 'Almaza'." *Krasnaya Zvezda*, 12 dek. 1992, 5; A. Rebrov, "Gonka za mirazhami, ili kuda vedet 'raketnyi sled'." *Krasnaya Zvezda*, 5 marta 1994, 5).
34 U. Aldrecht, *The Soviet Armament Industry*. Chur, Switzerland, Harwood Academic Publishers, 1993. (Originally published in German in 1989 as *Die sowjetische Rustungsindustrie* by Westduetscher Verlag, Opladen, Germany).
35 S.V. Kuvshinov and D.A. Sobolev, "Ob uchastii nemetskikj aviakonstruktorov v sozdanii reaktivnykh samolyotov SSSR." *Voprosy Istorii Estestvoznaniya i Tekhniki*, No. 1, 1995g, 103–114.
36 N.S. Stroev, "Ch. 6. Voennaya aviatsiya." in A.V. Minaev (Ed.), *Sovetskaya Voennaya Mosh' ot Stalina do Gorbacheva*. Moscow, Voennyi parad, 1999, 271–315.
37 N.S. Simonov, *Voenno-Promyshlennyi Kompleks SSSR v 1920–1950-e gody: Tempy Ekonomicheskogo Rosta, Struktura, Organizatsiya Proizvodstva i Upravlenie*. Moscow, ROSSPEN, 1999g, 227–240.
38 Analyzing the development of computers in the United States, Paul E. Ceruzzi suggests that the term "military" is misleading. He asserts that "there is no single military entity but rather a group of services and bureaus that are often at odds with one another over roles, missions, and funding." (Ceruzzi, *op. cit.*, in note 4. 7).
39 For an early perspective on this, see, for example, M. Boretsky, "Comparative Progress in Technology, Productivity and Economic Efficiency; U.S.S.R. Versus U.S.A." *U.S. Congress Joint Committee, New Directions in the Soviet Economy*, 1966, 172–175. M. Cave, "Computer Technology." in R. Amann, J.M. Cooper and R.W. Davies (Ed.), *The Technological Level of Soviet Industry*. New Haven, Yale University Press, 1977, 377–390. A.C. Sutton, *Western Technology and Soviet Economic Development 1945 to 1965*. Stanford, California, Hoover Institution Press, 1973, 319–323. Sutton suggests (362, 327) that the Soviet backwardness in this field may be accounted by the fact that Germany, from which the Soviet Union acquired most of its advanced technology, did not have much to offer at the time. However, we must take into consideration the fact that the Soviet Union also succeeded in certain fields, for which crucial know-how could not be gained from Germany, such as the atomic bomb.
40 Slava Gerovitch, *From Newspeak to Cyberspeak: A History of Soviet Cybernetics*. Cambridge, MA, London, The MIT Press, 2002, 7.
41 Ibid., 7, 8.
42 Ibid., 131.
43 See, for example, M.V. Keldysh, "Matematika – Vychislitel'naya tekhnika." in *Bol'shaya Sovetskaya Entsikropediya* (2-oe izd.), Vol. 50, 438. Moscow, Sovietskaya Entsiklopedia (Soviet Encyclopedia). I believe, however, that this dating *post factum* was framed up to insist on the advantage of BESM over another first Soviet computer in rivalry with BESM. In 1952, the development of the BESM-1 was far from completion.

44 RGAE, F. 8123, Op. 8, D. 560, 239.
45 Ibid., 204.
46 B.N. Malinovskii, *Istoriya Vychislitel'noi Tekhniki v Litsakh* Kiev, 1995. Electronic Version-Chap. 6. http://lib.ru/MEMUARY/MALINOWSKIJ/7.htm, 21.
47 I.M. Makarova et al. (Ed.), *Istoriya Informatiki v Rossii: Uchenye i Ikh Shkoly.* Moscow, Nauka, 2003, 186.
48 Virtual'nyi komp'yuternyi muzei, "Istoriya Otechestvennoi Vychislitel'noi Tekhniki." http://compmus9/valuehost.ru/histussr/18htm, 1.
49 S.P. Prokhorov, "Computers in Russia: Science, Education and Industry." *IEEE Annals of the History of Computing*, Vol. 21, No. 3, July–September 1999, 4–15.
50 G.D. Crowe and S.E. Goodman, "S.A. Levedev and the Birth of Soviet Computing." *IEEE Annals of the History of Computing*, Vol. 16, No. 1, Spring 1994, 4–24.
51 Gerovitch, *op. cit.*, in note 42. 131–149.
52 Ibid., 148, 149: However, we know that in their joint name, Igor' Kurchatov and Boris Vannikov, who eventually took the side of the Institute of Precision Mechanics and Computing Technique (*Institut tochnoi mekhaniki i vychislichel'noi tekhniki*; ITMiVT), a letter was sent to one of the Vice Prime Ministers, Vyacheslav Malyshev, an army general who also played an important role in the initial development of the Soviet atomic bomb, demanding the prohibition of publishing papers written about Strela because of "an unnecessary citation" and another reason [Rossiiskii gosudarstvennyi arkhiv ekonomiki (RGAE), Fond (F.) 8123, Opis' (Op.) 8, Delo (D.) 623, 271, 272].
53 Besides these two organizations, we can refer to the effort made by Isaak S. Bruk in Energetics Institute of the Academy of Sciences (see, Malinovskii, *op. cit.*, Chap. 5, http://lib.ru/MEMUARY/MALINOWSKIJ/6.htm.). He was, however, merely engaged in the computer development as a secondary job or hobby. His Institute did not provide any support to him, so I believe we may omit his work here.
54 Paul R. Josephson, *Red Atom: Russia's Nuclear Power Program from Stalin to Today.* Pittsburgh, PA, University of Pittsburgh Press, 2005, 2.
55 Ibid.
56 Yuka Tsuchiya, "Aizenhauâ Seiken-ki niokeru Amerika Minkan Kigyô no Genshiryoku Hatsuden Jigyô heno Sannyû (On the Entry of American Private Enterprises into the Atomic Power Generating Business)." in Tetsurô Katô and Mitsuo Igawa (Ed.), *Genshiryoku to Reisen: Nihon to Ajia no Genpatsu Dônyû* (*Atomic Energy and the Cold War: The Introduction of Atomic Power Stations to Japan and Asia*). Tokyo, Kaden-sha, 2013, 69.
57 The texts of the presentations read there were later published in a five-volume proceeding; Akademiya Nauk SSSR, *Sessiya po Mirnomu Ispol'zovaniyu Atomnoi Energii* (The Academy of Sciences of the USSR, *The Session on the Peaceful Use of Atomic Energy*), 1–5 Iyulya 1955g, Vol. 1–5. Moskva, Izd-vo AN SSSR, 1955.
58 Holloway, *op. cit.*, in note 9. 352: In addition to such a prominent scientist's ambition to earn international recognition, an outstanding historian of science in today's Russia, Vladimir Vizgin (1936–), suggests the significance of their intention to keep civil nuclear development apart from direct governmental control, as was seen in the case of military nuclear development, and put it under their own control. See, Vizgin, *op. cit.*, in note 16. 398–407.
59 John Krige, "Atoms for Peace, Scientific Internationalism, and Scientific Intelligence." *Osiris*, Vol. 21, No. 1, 2006, 167.
60 Ibid., 179.
61 Paul Josephson shows the details of that reactor (*op. cit.*, in note 54. 25–28).
62 Maika Nakao, Takeshi Kurihara and Masakatsu Yamazaki, "Resisting Nuclear Peril and Embracing Nuclear Power: Yasushi Nishiwaki, the Bikini Incident and

His Life in the Nuclear Age." *The Japan History of Science Society Historia Scientiarum*, Vol. 25, No. 2, August 2015, 8–35.

63 A.V. Lebedinskii (Ed.), *Sovetskie Uchyonye ob Opasnosti Ispytanii Yadernogo Oruzhiya*. Moscow, Atomizdat 1959g; Andrei Lebedinskii (1902–1965) says in the opening of this book, "A number of scientists failed to overcome the political pressures, which are given by the United States policymakers. Therefore, the scientists are taking some biased approach to the solution of problems. As a result, some biological disciplines even became the battlefields."

64 A biochemist, Norair Sisakyan (1907–1966: Later, Chief Academic Secretary of the Academy of Sciences), who attended Nishiwaki in his visit to the Soviet Union and was dispatched to Japan soon after Nishiwaki's return home, testifies in a Presidium meeting of the Academy on the influence of some Japanese scientists in this regard, saying, "Last fall, during my stay in Japan, I had a chance to talk with many Japanese radiobiologists who are extremely concerned about the presence of high radioactive contamination of some food products, for example, in rice, which is a staple of the diet for almost all the population in the East. The content of strontium rises to 260 strontium units" [Arkhiv Rossiskoi Akademii Nauk (ARAN), Fond (F.) 2106, Opis' (Op.) 1, No. 22, 2].

65 Toshihiro Higuchi, "'Kaku niyoru Heiwa' ni Chikyû-Kankyô-teki Genkai ha Aruka?: Hôshasei-Kôkabutsu no Anzensei-Shingi-Katei to Anzenhoshô-Kokka Amerika no Chiteki-Hegemonî no Kôzô to Hen-yô (Is There a Global Environmental Limit in 'Nuclear Peace'? The Japan Association for the International Relationship)." *Kokusai Seiji* (*International Politics*), No. 163, November 2011, 35 (in Japanese).

66 Hiroshi Ichikawa, "Against Lysenkoites' Hegemony: On the Establishment of the Institute of Cytology and Genetics at the Siberian Branch of the USSR Academy of Sciences." *Studies in the History of Biology*, Vol. 9, No. 2, 2017, 7–21: This paper is now a part of this book (Chapter 7).

67 Valerii Soifer, *Vlast' i Nauka: Razgrom Kommunistami Genetiki v SSSR* (*Political Power and the Science: A Pogrom of Genetics by the Communists in the Soviet Union*). Moscow, Izd-vo, CheRo, 2002r, 829–835, 860.

68 Ibid., 607–608.

69 See, Kirill Rossiyanov, "1948-nen Zenrenpô Nôgyô Kagaku Akademî 8-gatsu Sôkai niokeru Ruisenko-ha no Shôri (The Lysenkoites' Victory at 'August Session' of All-Union Academy of Agricultural Sciences Named After V.I. Lenin)." in Hiroshi Ichikawa (Ed.), *Kagaku no Sanbô-honbu: Rosia/Sorenpô Kagaku Akademî ni kansuru Kokusai Kyôdô Knekyû* (*The General Staff of Science: An International Joint Study on Russian/Soviet Academy of Sciences*). Sapporo, Japan, Hokkaido University Press, 2016, 313–329 (in Japanese).

70 Soifer, *op. cit.*, in note 69. 860.

71 Paul Josephson, *New Atlantis Revisited: Akademgorodok, the Siberian City of Science*. Princeton and Oxford, Princeton University Press, 1997, 95–108; S.V. Shalimov, "'Spasenie i vozrozhdenie': istoricheskii ocherk razvitiya genetiki v Novosibirskom nauchnom tsentre v gody 'ottepeli'." *Istoriko-biologicheskie issledovaniya* (*Studies in the History of Biology*), Vol. 3, No. 4, 2011, 136–138; S.V. Shalimov, "Razvitie genetiki v Novosibirskom nauchnom tsentre vo vtoroi polovine 1960-kh. gg.: Sotsial'no-istoricheskii aspekt." *Istoriko-biologicheskie issledovaniya* (*Studies in the History of Biology*), Vol. 5, No. 1, 2013, 16–32, and his other writings. See also, the site of this Institute, www.bionet.nsc.ru/.

72 Soifer, *op. cit.*, in note 69. 848.

73 Slava Gerovitch, *op. cit.*, in note 40. 183, 184. The original text of the "Letter of Three Hundreds" can be seen in A.F. Kiselev and E.M. Shchagin (Ed.), *Khrestomatiya po Otechestvennoi Istorii (1946–1995)*. Moscow, Izd-vo 'Vlados', 1996, 458–460; Otherwise, Paul Josephson refers also to the important role

of the direct tie between some anti-Lysenko biologists like Nikolai Timofeef-Ressovskii (1900–1981) and the leading physicists like Igor' Tamm (1895–1971. A Nobel laureate in Physics in 1958) (Josephson, *op. cit.*, in note 54. 87–90). Then, Soifer suggests the relationship between the scientists' call for "normalization of biology" and the importance of the study on "radiation effects on living body" (Soifer, *op. cit.*, in note 69. 64–867). However, his view is apt to be concentrated in personalities of some leading scientists and to lose the organizational context.

74 A.V. Lebedinskii (Ed.), *op. cit.*, in note 63. The book comprises the following chapters: "In the place of a preface," "The harmful effects of nuclear weapons testing," "On E. Teller and A. Latter's *Our Nuclear Future: Facts, Dangers, and Opportunities*," "Radiocarbon generated by nuclear explosions and the non-threshold biological effects," "The contamination of the biosphere around Leningrad by radioactive matters created by the nuclear bomb explosions," "Absorption of radioactive strontium by plants and its accumulation in various agricultural products," "A study on the internal containment of radioactive strontium in air, soil, food, and human bones," "Radiation and human genes," "The cytogenetic consequences of radiation effects in the spermatozoa formation of the apes," and "The effects of radioactive strontium on embryonic heredity." Most of these chapters report the results of prior observations and experiments. Excluding "On E. Teller and A. Latter's *Our Nuclear Future: Facts, Dangers and Opportunities*" (author: Ovsei Leipunskii, 1909–1990), which took a book-review style, the chapters that borrowed data from Western scientists are the following: "Radiation and human genes" (author: Nikolai Dubinin), which reports findings from a number of Western geneticists, such as Hermann Joseph Muller (1907–1998. A Nobel Prize laureate in Physiology or Medicine in 1946); "Radiocarbon generated by nuclear explosions and the non-threshold biological effects" (author: Andrei Sakharov), which discusses the work of Ernest Anderson (1920–2013) and sets out an original calculation method; and "The harmful effects of nuclear weapons testing" (Andrei Lebedinskii), which presents a general outline of nuclear weapons testing and discusses research papers by the U.S. Health and Safety Laboratory and articles published in the journal *Science*. Thus, the scientists did not use any data that had come from Japanese sources, except for the Japanese data submitted to UNSCEAR.

75 M.N. Pobedinskii, *Luchevaya Bolezn'* (*Posledstviya Porazheniya Organizma pri Atomnom Vzryve*). Moskva, Medgiz, 1957g.

Part I

What drove the scientists to the Cold War?

In the case of physicists

1 "Jealousy" and "discord"

The conflicts among physicists during the war

This chapter aims to shed new light on the material conditions that surrounded Soviet physicists, who were later involved in the ideological dispute in the second half of the 1940s, focusing the differences between the full-time university staff and the part-time staff who worked in concurrence with other jobs they had mainly at the Physics Institute named after P.N. Levedev of the Soviet Academy of Sciences.

1 The gaps in research conditions between "two categories" of physicists: Faculty of Physics of Moscow State University and the Physics Institute of the Academy of Sciences in the prewar period

Soviet science was able in several areas to achieve great results and stay at the forefront of the world scientific scene. Without any doubt, the most striking feature of Soviet science consists in its organizational framework in which the Academy of Sciences played a crucial role in the historical development; Soviet universities and other institutions of higher education were given the specialized task of providing education.[1] In West-European countries such as France and the United Kingdom "academies of science" had played a major role in the formation and development of the profession of "scientist," shaping its social identity and status. Ultimately, however, it has lost its original function and became honorary an organization for acclaimed scientists.[2] The Soviet (and, in part, Imperial Russian) Academy of Sciences, however, consistently dominated in the scientific world as an independent, powerful, and practical organization, exercising total control over research activities throughout the country by means of holding advanced research facilities and the expertise under its umbrella.[3]

Until 1934, however, when most of the facilities of the Academy of Sciences were still located in Leningrad, the separation of research and education was not so clear-cut. In 1922, the staff of the Faculty of Physics (*Fizicheskii fakul'tet*. Hereafter, Fizfak) of Moscow State University (Hereafter, MSU), then the Faculty of Physics and Mathematics, had founded the Scientific Research Institute of Physics and Crystallography, modeling it on the Physics Institute of MSU, founded in 1904. It was to serve as a base for Fizfak staff's research activities.

The Institute was renamed the Scientific Research Institute of Physics (*Nauchno-issledovatel'skii institut fiziki*, or NIIF) in 1927.[4] According to a report submitted in 1927 by V.I. Romanov (1880–1954), then Director of NIIF, his institute was composed of ten regular members, of which six were not the regular staff of MSU; totally 16 staff, including the regular staff, worked at the institute, of which more than half did not belong to MSU. Although this institute formally belonged to MSU, it actually consisted of the physicians from various institutes of higher education and industrial institutes from all around Moscow.[5]

As a consequence of reform in 1930, led by the then Director of NIIF, and, at the same time, Dean of Fizfak, Boris Gessen (1893–1936), earlier than the overall reform of the higher education related with the unsuccessful and confused consequence of "Cultural Revolution," NIIF was expanded to a large-scale, genuinely powerful research institute.[6] The number of its personnel was increased to approximately 80 persons, including 20 highly qualified leaders. NIIF then was comprised of a theoretical division and seven laboratories: optics, physics of heat, X-ray structural analysis, magnetism, electrical phenomena in gas, oscillation, and short wavelength.[7] It is worthy of mention that in the laboratory of optics Grigorii Landsberg (1890–1957) and Leonid Mandel'shtam (1879–1944) discovered the inelastic scattering of a photon, i.e., the so-called Raman Effect, in 1928 independently of Chandrasekhara Venkata Raman (1888–1970. A Nobel Prize laureate in Physics in 1930) and Kariamanickam Srinivasa Krishnan (1898–1961. A Nobel Prize laureate in Physics in 1930).[8] Landsberg also developed a method of high-speed, quantitative spectral analysis.[9] In the laboratory of magnetism a method for magnetic structural analysis was developed by Nikolai Akulov (1900–1976), who discovered the law of induced anisotropy in Fizfak in 1928.[10] In the laboratory of oscillation, Leonid Mandel'shtam and Nikolai Papaleksi (1880–1947. Then he worked in Leningrad) were dealing with the problem of self-excited oscillation in cooperation with some young physicists such as Aleksandr Andronov (1901–1952), Aleksandr Witt (1902–1938), and Semyon Khaikin (1901–1968).[11] In the theoretical division, Igor' Tamm (1895–1971. A Nobel Prize laureate in Physics in 1958) and Yurii Rumer (1901–1985) were struggling to develop solutions to general problems of quantum mechanics, while Mikhail Leontovich (1903–1981) engaged with the principle problems of statistical physics. Their work attracted the attention of foreign physicists, and they were sometimes referred to as the "Moscow School" abroad.[12]

Also in 1933, the leaders of NIIF continued to develop the institute. During the Second Five Year Plan (1933–1937), NIIF was regarded as the "leading" research institution in physics in Moscow, while the Physical Technology Institute, headed by Abram Ioffe (1880–1960), was regarded as the "leading" one in Leningrad. As early as in 1934, however, NIIF found itself being encroached upon; some outstanding physicists were recruited by the Academy's Physics Institute (*Fizicheskii Institut im. P.N. Levedeva Akademii nauk SSSR*. Hereafter, FIAN), which was separated from the Institute of Physics and Mathematics of the Academy of Sciences and established as an independent research institute in Moscow in that year. On October 15, 1934, Tamm, Landsberg, Leontovich and Sergey Rytov (1908–1996) were hired by FIAN at a wage of 500 rubles (for Tamm and Landsberg), 450 (for Leontovich), and 350 (for Rytov). In the following year, the Director of NIIF, Gessen, was recruited by FIAN as a senior expert, and then as Deputy Director. In November 1935, the Dean of Fizfak, Khaikin was recruited by FIAN. Following Khaikin's appointment, the most influential figure at Fizfak and NIIF, Mandel'shtam, was also recruited to FIAN at a wage of 750 rubles. They continued their work at Fizfak and NIIF alongside their research activities at FIAN. In the same year Papaleksi moved to Moscow in order to work for FIAN. Tamm, Landsberg, Leontovich, and others who began work at FIAN remained also at NIIF, but in name only. In 1934, Aleksandr Andronov (1901–1952) left Moscow to work in Gor'kii, and soon after, Gabriel' Gorelik (1906–1957) followed him. Witt and Sergei Strelkov remained at NIIF but not for too long. In 1937, Witt was arrested and later died in a prison camp. NIIF was on the verge of being deprived of its talented staff.[13]

In spite of this exhaustion and some disorders having followed the arrest of Gessen, NIIF continued its activity thanks to the efforts of the remaining staff of NIIF and Fizfak, some of whom succeeded in procuring research expenses in trust from several industrial organizations and other benefactors.[14] On the basis of the abstract prepared for a lecture in 1938, the very influential monograph by Akulov, titled *Ferromagneitsm*, was published under the edition by Konstantin Belov (1911–2001).[15] Published in *Zhurnal Eksperimental'nykh i Teoreticheskikh fizik* (*Journal of Experimental and Theoretical Physics* (Hereafter, *ZhETF*), the paper "On Vibration Properties of the Electron Gas" prepared by Anatolii Vlasov (1908–1975) received international fame later. The paper gave the first thorough analysis of the physical properties of the charged particles (plasma) that illustrated the inapplicability of the Boltzmann gas-kinetic equation and proposed a new kinetic equation that took into account the collective interaction between charged particles.[16]

The actual amalgamation of two formerly independent structures, Fizfak and NIIF, went ahead because of three major factors: the increase in concurrent works among virtually all research and teaching staff at NIIF and Fizfak; the fact that the recently graduated students from Fizfak began to work at NIIF (or in Fizfak); and the fact that, from 1937, the roles of Director of NIIF and Dean of Fizfak were performed by one person.[17] It is under these circumstances that, at a meeting at NIIF on October 16, 1938, Akulov said,

> We have a number of departments with relevant research laboratories and relevant scientific staff. However, there are, in addition, a number of subsidiary organizations – the works, the drafting room, the warehouse and a number of ancillary scientific and technical workers attached to the Institute. In this way, – it turns out that the departments have everything necessary for conducting the technical processes. These same staff members are also carrying out research work. Ensuring research work, however, depends on the staff of the Institute. So this proves that the Institute plays only an ancillary, technical role.[18]

In other words, Fizfak had been gradually taking the function of scientific research activities. In 1940, a total of 54 postgraduates studied at NIIF.[19]

2 Wartime evacuation of the Faculty of Physics and return to Moscow

After German troops entered the Soviet Union in June 1941, the main part of MSU began to be evacuated to Ashkhabad in accordance with the resolution of the State Defense Committee dated October 15, 1941.[20] Without a delay 68 professors, 58 associate professors, and 39 assistants moved to this city and began to teach the students in a normal school building under construction.[21] With a turn for the better in the tide of the war, as early as June 1942, MSU began to move from Ashkhabad to Sverdlovsk (now Ekaterinburg), but before all of MSU could be transferred to Sverdlovsk, the government decided to move MSU to Moscow; those members who had already moved to Sverdlovsk remained there only six months. Several faculties came directly from Ashkhabat to Moscow. The return to Moscow was completed on June 10, 1943.[22] As for Fizfak, its main part was evacuated first to Ashkhabat, then to Sverdlovsk, but a small group, headed by Boris Il'in (1888–1964), remained in Moscow and continued to work under harsh conditions. They accommodated the returning staff and students and conducted wartime research, mainly at NIIF; for example, Khaikin was engaged in the radio wave problem related to radar research and development in the underground floor of Fizfak (although his research was soon transferred to FIAN with the participation of Leontovich, who returned to Moscow earlier than the other FIAN staff). Fyodor Korolyov (1909–1979) was awarded a state prize for developing a method for high-speed spectrum analysis of metal and a method for super high-speed photography of intertwined gas currents which he made in Moscow during the war. Following his return to Moscow, the Dean of Fizfak and Director of NIIF Aleksandr Predvoditelev (1891–1973) engaged with the problem of ignition of gaseous mixtures in high temperatures. In spite of such great efforts made by the staff of NIIF, their research became impossible to do by 1943, when, in addition to the requisition made in 1942, many more machines and equipments required by NIIF were redesigned to fulfill orders from the front.[23]

And after the end of the war, on January 28, 1946, the government ordered the installation of a special research facility for nuclear physics within MSU. Thus, the Second Scientific Research Institute of Physics (NIIF-2) was established with Dmitrii Skobel'tsyn (1892–1990) as Director. The institute was to have a 4MeV-class cyclotron. NIIF, now renamed NIIF-1, remained the base for "patriotic and materialistic physicists." Later, in 1953–1954, NIIF-1 was heavily criticized for its insufficient performance, and a governmental commission in charge of inspection of the scientific and educational activities of MSU, which was formed after "the students' speeches in 1953" incident, recommended its closure because of its backwardness in equipment and methods.[24]

There is not much information available about the works and lives of Fizfak staff in Ashkhabad. According to the review of Professor Roman Telesnin

(1905–1985), during the evacuation, almost all research activities stalled.[25] In Sverdlovsk, Fizfak was housed in the building of the Ural Polytechnic Institute (Figure 1.1). During the war, however, that institute actually turned into a factory for artillery weapons and ammunition, and the halls and classrooms were filled with tools and machines. Only in the evenings, so often in the hallways, were lectures given to the students.[26] In such a situation, the professors and teachers had to give up any thoughts of conducting their own research.

After the resettlement in Moscow, Fizfak staff experienced considerable difficulty because of lack of space. MSU as a whole expanded and became a vital institute for the training the experts in various fields of Soviet society. By the end of the war, it had established four new faculties, such as the Economics Faculty, the Faculty of Law, and others, and the number of departments (*kafedra*) had increased from 107 to 156.[27] During the war, the number of the full-time professors was increased to 284, part-time professors – 116, full-time associate professors (*dotsent*) – 262, part-time associate professors – 155, full-time lecturers – 267, and part-time lecturers – 110 as of October 1945.[28] It is worthy of note that MSU was strengthened to a remarkable extent during the war, in spite of the confused situation and the withdrawal of a large number of the students from the university for the confused situation, military service, and other reasons.[29] Fizfak accepted 341 new students in 1945,[30] in contrast to 150 students in 1941.[31] Fizfak established the Branch for Geography in 1943[32]; the Department of Low Temperature Physics headed by Pyotr Kapitsa (1894–1984. A Nobel Prize laureate in 1978), who was then Director of the Institute of Physical Problems and was installed as Chief of the Main Directorate of Oxygen Industry, a governmental agency responsible for the supply of oxygen for reinforcement of explosive power during the war; the Department of Oceanic Physics[33]; and, in November 1944, the Department of Nuclear and Radiation Physics headed by Igor' Kurchatov (1903–1960), a leader of the Soviet Atomic Project, as a part-time professor.[34] In addition Fizfak absorbed some teachers while it was based in Sverdlovsk; therefore Dmitrii Ivanenko (1904–1994) and Arsenii Sokolov (1910–1986) from Sverdlovsk State University restarted their careers as MSU staff during the war.[35]

The then-Rector of MSU, Il'ya Galkin (1898–1990) repeatedly drew the attention of the government and the leadership of the city of Moscow to the lack of space, a particularly difficult situation.[36] A standard floor space for one student was estimated at 12 meters square as the rule. However, in reality only less than half of the standard space was secured. The newly established faculties and departments were not allotted any space.[37] Even in 1947, Fizfak was allotted 4,800 m^2, in which only 2,681 m^2 could be utilized for the education and research activities. Fizfak staff were forced to use the spaces of the research facilities where they worked concurrently with MSU, the corridors, and other spaces. Due to such shortage of the educational space, Fizfak often had to reduce the mandatory training hours (lectures, seminars, workshops, etc.) and had to conduct them on a three-shift system from 9 o'clock in the morning until 11 o'clock at night.[38] In addition, "studying by oneself became a remarkable part in a learning plan."[39]

Figure 1.1 The hall of Ural Polytechnic Institute (now, University), where Fizfak was evacuated to. Then, this hall was filled with machine tools for munition manufacture.

Source: The collection of the author.

3 The evacuation of the Academy's Physics Institute to Kazan

After the Germans invaded the Soviet Union, the institutes of the Academy of Sciences of the USSR located in Moscow and Leningrad were evacuated on an unprecedented scale to Kazan and other Soviet cities.[40] The staff of FIAN began to move to Kazan on July 22, 1941. They arrived there on the 25th with a large amount of equipment. All the equipment was installed at its new location in just a week and immediately began to be used for scientific research for military purposes. Altogether, 45 of the 50 scientific staff of FIAN were evacuated to Kazan.[41] There, they suffered from a lack of space and shortage of food. The stage and gymnastic hall of the main building of the Kazan State University (Figure 1.2), into which many institutes of the Academy of Sciences were evacuated, became a "dormitory" for 200 persons (Figure 1.3). In addition, in private homes, two families were often settled, in the same room, dividing one room with a partition of sheets.[42] Thus, for example, N. Sobolev, a FIAN staff member, had only 6 m^2 for himself, his wife, and their son in a room, shared with another family. Officially, 600 gram of bread had to be distributed per person per day, but this rate was very often reduced to 400 gram. Besides bread, in the dining room the staff could get a bowl of pea soup and a pea liquid porridge, which sometimes included potato cutlets.[43]

Figure 1.2 The main building of Kazan State University, where FIAN was evacuated to.

Source: The collection of the author.

Figure 1.3 The stage and gymnastic hall as a "dormitry." (A physicist, Yakob Frenkel, painted.)

Source: The collection of the author (a close-up picture of an exhibit at the Museum of History of the Kazan State University).

Nevertheless, the scientists in Kazan vigorously engaged in wartime research, which included the development of the method of quantitative analysis of toluene,[44] spectral analysis for the determination of the thickness of alloyed parts using gamma rays, the use of stereometry for X-ray diagnostics, the development of methods and devices against combat radio interference from local sources, the theory and practical patterns of diffusion of short and ultra-short radio waves, the development and introduction of germicidal lamps, the production of daylight lamps, the development of ceramic condensers for radios, and the spectral methods of analysis of military and industrial materials.[45]

What is important here is that many of FIAN's staff remained engaged in their military-related tasks even after their return to Moscow. For example, Leontovich and Lev Artsimovich (1909–1973) were called to give lectures at a secret educational facility named the "Physics Engineering Faculty" of MSU, which was established to train engineers for the development and production of the "newest" weapons, as its forerunner, Moscow Mechanical Institute of the People's Commissariat for Ammunition had done previously.[46] Thus, many Fizfak professors and teachers working part time in concurrence with their jobs at FIAN were very often unable to commute to MSU in spite of their official obligation.[47]

4 The accusation of Fizfak professors

At the beginning of 1944, when almost all Fizfak staff had returned to Moscow, but when many of its part-time teachers belonging to FIAN were unable to commute to MSU, an election was held for the new Chair of the Fizfak's Department of Theoretical Physics. Vlasov was chosen as the new Chair, obtaining 24 votes against Tamm's five. This result of the voting shows the hate of the majority of the full-time teachers in Fizfak against several part-time teachers. According to the reminiscences of Yakov Terletskii (1912–1993), the Dean, Predvoditelev did not expect the returns to the positions in MSU of Leontovich and Tamm who didn't come back to the university from the evacuation, regarding their attitude as that of those who "abandoned the Faculty of Physics."[48] Khaikin, who had stayed in Moscow during the war, however, but then mainly worked in FIAN, was dismissed from MSU. Landsberg was also refused his return to MSU.[49] Mandel'shtam died in the end of that year.

On the other hand, although Vlasov was a relatively young, talented physicist, from the viewpoint of Vladimir Fok (1898–1974), he was a troublesome person who "fanatically believe in an incorrect idea, an idea that the method he uses for the solution of theory on plasma has universality."[50] In support of Tamm, four renowned physicists, Abram Ioffe, Abram Alikhanov (1904–1970), Aleksei Krylov (1863–1945), and Kapitsa sent a joint petition to Vice-Primier Vyacheslav Molotov (1890–1986) in July 1944 to prevent Vlasov's assumption of the Chair of the department. They were successful. The government's Committee for the Affairs on Higher Education, with Sergei Kaftanov (1910–1978) in the Chair, directly and irrevocably appointed a professor from Leningrad State University, Fok to the position. This incident made tangible an internal conflict between two groups of physicists within the country.[51] However, this high-handed measure made the "rifts" within the Faculty of Physics more serious.

In the middle of that incident, in the spring of 1944, five professors and a teacher from Fizfak – Predvoditelev, Akulov, Nikolai Kobozev (1903–1974, a chemist), Nikolai Zelinskii (1861–1955, a chemist), Arkadii Timiryazev (1880–1955, a philosopher than a physicist), and Vasilii Nozdryov (1913–1995) – simultaneously sent letters to Secretary of the Central Committee and, at the same time, the First Secretary of Moscow Committee of All-union Communist Party, Aleksandr Shcherbakov (1901–1945), making a number of accusations against leading scientists, including the now late Mandel'shtam, Landsberg, Tamm, a physical chemist Aleksandr Frumkin (1895–1976), and a chemist Nikolai Semyonov (1896–1986. A Nobel Prize laureate in Chemistry in 1956); Timiryazev accused Tamm on the basis of his intimacy with a Trotskyist, Gessen, and Akulov accused Semyonov on the basis of his "concoction" of scientific works. The main thrust of their accusation, however, focused on the accuseds' "monopoly" of academic posts, laboratory floor spaces, and the spaces for scientific papers in various scientific journals and elsewhere.[52]

Their accusations were not accepted by the political authorities, however; Kaftanov, then Chair of the Committee for the Affairs on Higher Education, in charge of the case, sent a report on April 27, 1944, to Politburo member and Vice-Primier, Georgii Malen'kov (1902–1988), stating that the accusers "raised a question in a proper way that scientific problems must be developed also at universities, especially at MSU," but "their accusation is not acceptable" because of the importance of the scientific research and training of new specialists by the accused scientists.[53] A further, a letter, dated September 8 and jointly signed by Kaftanov, Mikhail Pervukhin (1904–1978, then Minister of Chemical Industry, confidentially in charge of nuclear development), Sergei Suvorov (1902–1994. An executive in the Propaganda and Agitation Section of the Party's Central Committee), and Nikolai Bruevich (1896–1987. Academician-Secretary of Presidium of the Academy of Sciences) was sent to Malen'kov and Shcherbakov, reporting the results of the reexamination of this case. They revealed the groundlessness of every accusation, concluding that "they turned not to be regarded leading scientists in the Soviet Union, owing to the result of their own activities."[54]

Notes

1 Loren R. Graham, *Science in Russia and the Soviet Union: A Short History*. Cambridge, Cambridge University Press, 1993.

2 See, Sayaka Oki, "Kagakusha ha Itsu-kara Sonzai shite-ita no-ka? (When Did Scientists Appear?)." in Michiyo Nakane et al. (Ed.), *Kagaku no Shinri ha Eien-ni Huhen na-no-darô ka? (Will Scientific Truth Be Everlasting?)*. Tokyo, Publisher Bere, 2009, 103–128.

3 Bearing in mind the role played by the Academy of Sciences in the history of the Soviet Union, its organizational features and the socio-political environment in which it operated should be an important subject for the history of science, and of sociology of science. Aware of the importance of this research subject, six historians of Russian and Soviet science have been co-conducting a research project supported by a grant-in-aid from the Japan Society for the Promotion of Science for 2010–2012 fiscal years. The research result of this project was crystallized in an international collective work, Hiroshi Ichikawa (Ed.), *Kagaku no Sanbô-honbu: Rosia/Sorenpô Kagaku Akademî ni kansuru Kokusai Kyôdô Knekyû (The General Staff of Science: An International Joint Study on Russian/Soviet Academy of Sciences)*. Sapporo, Japan: Hokkaido University Press, 2016.

4 A.V. Andreev, *Fiziki Ne Shutyat: Stranitsy Sotsial'noi Istorii Nauchno-Issledovatel'skogo Instituta Fiziki pri MGU (1922–1954)*. Moscow, Progress-Traditsiya, 2000, 18, 19.

5 A.A. Pechyonkin, *Leonid Isaakovich Mandel'shtam: Issledovanie, Prepodavanie i Ostal'naya Zhizn'*. Moscow, Logos, 2011, 240.

6 Narodnyi Kommisariat Prosveshcheniya RSFSR, *Universitety i Nauchnye Uchrezhdeniya k XVII S"ezdu VKP(b)*. Moscow, Leningrad, Gosudarstvennoe Tekhniko-Teoreticheskoe Izdatel'stvo, 1934, 21, 22; In the midst of the "Cultural Revolution," in November 1929, the Party' Central Committee adopted a policy of separating some faculties from the universities and transferring them to the independent institutes of higher education. The universities became specialized institutions for mathematics and some branches of natural sciences. In addition, the use of the "brigade-style" in laboratory works, introduced as a reaction to the rigidity of the teaching methods in higher education and for the sake of quick supply of specialists needed for the socialist construction, reduced the learning time

in every step of university education from the Faculty for the Workers to the post-graduate studies and, thus, led to the decline of academic standards and down-grading the level of graduates. Consequently, as early as July 1931, the Ministry of Education of the Russian Soviet Federal Socialist Republic adopted a resolution for an orbit adjustment. One of the measurements taken by the Ministry was the establishment of scientific research institutes affiliated to universities for the sake of transformation of universities from mere educational facilities to scientific and educational facilities (Narodnyi Kommisariat Prosveshcheniya RSFSR, *Gosudarstvennye Universitety.* Moscow, Ogiz-Izogiz, 1934, 18).

7 *Universitety i Nauchnye Uchrezhdeniya . . . op. cit.*, in note 6. 34.

8 *Gosudarstvennye Universitety . . . op. cit.*, in note 6. 111; The original edition is – G. Landsberg and L. Mandelstam, "Eine neue Erscheinung bei der Lichtzerstreuung in Krystallen." *Naturwissenschaften*, Vol. 16, No. 28, 1928, 557, 558.

9 *Universitety i Nauchnye Uchrezhdeniya . . . op. cit.*, in note 6. 35.

10 *Gosudarstvennye Universitety . . . op. cit.*, in note 6. 111.

11 Ibid., 111.

12 *Universitety i Nauchnye Uchrezhdeniya . . . op. cit.* 41.

13 Pechyonkin, *op. cit.*, in note 5. 240–243.

14 Ibid., 243.

15 N.S. Perov, *Nikolai Sergeevich Akulov.* Moscow, Fizicheskii Fakul'tet MGU, 2003, 29.

16 I.P. Bazarov and P.N. Nikolaev, *Anatolii Aleksandrovich Vlasov II.* Moscow, Fizicheskii Fakul'tet MGU, 1999, 10; This work of Vlasov later became the target of criticism from some leading physicists such as Vladimir Fok (1898–1974), Leontovich, Lev Landau (1908–1968. A Nobel Prize laureate in 1962) and Vitalii Ginzburg (1916–2009. A Nobel Prize laureate in 2003) (Andreev, *op. cit.*, in note 4. 104, 297, 298).

17 Andreev, *op. cit.*, in note 4. 97.

18 Ibid., 9, 98.

19 Tsentral'nyi Munitsipal'nyi Arkhiv Moskvy (TsMAM), Fond (F.) 1609, Opis' (Op.) 2, Delo (D.) 134, 7.

20 A.S. Ilyushin (Ed.), *Vasilii Stepanovich Fursov.* Moscow, Fizicheskii Fakul'tet MGU, 2010, 56.

21 TsMAM, F. 1609, Op. 2, D. 134, 7.

22 Andreev, *op. cit.*, in note 4. 100.

23 Ibid., 99, 100; What is miserable concerning this mobilization of the machines and equipment from NIIF is a cry of Fizfak professor, Arkadii Timiryazev (1880–1955), in a Party meeting of January 3, 1944, "The communists must be able to prepare their own dissertations. The defenders of a thesis must be able to prove the thesis through experiments . . . " (Tsentral'nyi Arkhiv Obshchestvenno-Politicheskoi Istorii Moskvy, F. 478, Op. 1, D. 75, 14).

24 Yu.V. Goponov, S.K. Kovalyova and A.V. Keccenikh, "Studencheskie vystupleniya 1953 goda na Fizfake MGU kak sotsial'noe echo atomnogo proekta." in V.P. Vizgin (Ed.), *Istoriya Sovetskogo Yadernogo Proekta: Dokumenty, Vospominaniya, Issledovaniya*, Vol. 2. St. Petersburg, Izd-vo Russkogo Khristianskogo Gumanitarnogo Instituta, 2002, 521, 522, 527–537; Andreev, *op. cit.*, in note 4. 147–153.

25 S.Kh. Karpenkov, *Roman Vladimirovich Telesnin.* Moscow, Logos, 2004, 16.

26 Interview with the Director of the Museum of the Ural State Polytechnic University-Ural Polytechnic Institute, Yu.B. Shaton, at the Ural State Polytechnic University on February 18, 2008.

27 TsMAM, F. 1609, Op. 2, D. 194, 1, 7: A Russian word *Kafedra* means a unit of university professors and teachers classified along their specialties.

28 TsMAM, F. 1609, Op. 2, D. 183, 4.

29 For example, in 1945, totally 572 students were expelled from the university, of which 11 were expelled for military service, 99 – for their lack of the precondition for attendance in the university and the others – for their poor performance.
30 TsMAM, F. 1609, Op. 2, D. 183, 1.
31 TsMAM, F. 1609, Op. 2, D. 126, 7: In this document, a part-time teacher was also noted as "a teacher provided half salary."
32 TsMAM, F. 1609, Op. 2, D. 154, 14.
33 TsMAM, F. 1609, Op. 2, D. 194, 1.
34 M.I. Panasyuk, E.A. Romanovskii and A.V. Kessenikh, "Nachal'nyi etap podogotovki fizikov-yadershchikov v Moskovskom gosudarstvennom universitete (tridtsatye – pyatidesyatye gody)." in Vizgin (Ed.), *op. cit.*, in note 24. 501: In the early days of 1945, it was decided to establish the nuclear laboratory within the Department of Nuclear and Radiation Physics. Il'ya Frank, Sergei Vernov, and Vladimir Veksler, Sergei Vasil'ev (a specialist in particle accelerators) soon began to work there. This laboratory was allotted a new space, despite the overcrowded situation in MSU campus (ibid., 501, 502). After the war, in January 28, 1946, the government instructed MSU to establish a new research institute for atomic nuclear study. Dmitrii Skobel'tsyn was appointed Chief of the newly established institute, i.e., the Second Scientific Research Institute of Physics (NIIF-2) (ibid., 503). By 1949, a 4 MeV-class cyclotron was completed for the first time in educational facilities in the Soviet Union. In December 1948, the Department of Nuclear and Radiation Physics was upgraded to the Branche of Material Structure (later, Atomic Nucleus). The branch was composed of five departments: Atomic Nucleus, Accelerators, Neutron Physics, Nuclear Spectroscopes, and Cosmic Ray. The branch accepted more than 100 students every year (ibid., 505, 506). In 1957, NIIF-2 was renamed to the Scientific Research Institute of Nuclear Physics, Director of which was then Vernov, who succeeded the post after Skobel'tyn in 1946 on the occasion of his transfer to the United Nations. On the other hand, NIIF-1, the former NIIF, was under the sphere of influence of such professors as Akulov, M.D. Karasyov, Korolyov, Nozdryov, Terletskii, and other full-time teachers in Fizfak. Later, in 1949, they showed their intention to conduct their own nuclear study by establishing their own nuclear laboratory (Goponov et al. "Studencheskie vystpleniya 1953 goda . . ." in Vizgin (Ed.), *op. cit.*, in note 24. 521, 522).
35 Andreev, *op. cit.*, in note 4. 101.
36 TsMAM, F. 1609, Op. 2, D. 164, 19-b; TsMAM, F. 1609, Op. 2, D. 175, 8.
37 TsMAM, F. 1609, Op. 2, D. 194, 1, 7.
38 TsMAM, F. 1609, Op. 2, D. 219, 4: In addition, the overaged equipment also annoyed the teachers of Fizfak. For example, a meridian circle manufactured in 1847 (repaired in 1900) was also used even in 1947. Although approximately 300 boxes of equipment acquired in Germany as "trophies" were sent to the Faculty of Physics, almost all of them were useless; some of them lacked parts like amplifiers without vacuum tubes, others were overaged (TsMAM, F. 1609, Op. 2, D. 194, 9–11).
39 TsMAM, F. 1609, Op. 2, D. 164, 19-b: Even in such a situation, Fizfak teachers made great efforts to improve their educating methods. "Study from a clean desk," a unique training method to make students for themselves select, build up, and adjust the research instruments and devises suitable for the task given by teachers was invented by them in those days and highly evaluated then in MSU (TsMAM, F. 1609, Op. 2, D. 199, 18).
40 Except for the Institute of Crystallography established during the war, the evacuated institutes of the Academy of Sciences can be divided into four groups: (1) ten institutes, evacuated from Moscow to Kazan; (2) three institutes, evacuated

from Moscow to Sverdlovsk (along with nine boards, commissions, and other institutions); (3) seven institutes, evacuated from Leningrad to Kazan and Stalinabad; and (4) four institutes evacuated from Moscow to Frunze and other cities in Central Asia. Having examined the conditions under which the evacuated institutes were engaged in various activities in their wartime evacuation, the author became aware of the very big differences among them that might offer clues to solving several puzzles of the postwar history of Soviet science, such as the riddle of the antagonism shown by the great mathematician Mikhail Lavrent'ev (1900–1980) toward the nuclear physicists working at the Institute of Nuclear Physics of the Siberian Branch of the Academy of Sciences [See, Hiroshi Ichikawa, "Sorenpô Kagaku Akademî no Senji-Sokai ni kansuru Ichi Kôsatsu (On the Evacuation of Institutes of the USSR Academy of Sciences during the Second World War)." The Graduate School of Integrated Arts and Sciences of Hiroshima University, *Shakai Bunka Roshû* (*Hiroshima Papers on Society and Culture*), No. 11, March 2010, 1–28].

41 Arkhiv Rossiskoi Akademii Nauk (ARAN), F. 532, Op. 1, D. 69, 1.

42 M.S. Tagirov, B.G. Tarasov and S.V. Pisareva, *Fizicheskie Instituty Akademii Nauk SSSR v Kazanskom Universitete v Gody Velikoi Otechestvennoi Voiny.* Kazan', The Kazan' State University Press, 2005, 6.

43 N.N. Sobolev, "O Kazanskom Periode FIAN (1941–1943)." in *Fizicheskii Institut im. P.N. Lebedeva v Gody Velikoi Otechestvennoi Voiny.* Moscow, The Physics Institute Press, 1995, 25–27.

44 ARAN, F. 532, Op. 1, D. 64, 1.

45 ARAN, F. 532, Op. 1, D. 74, 1–4.

46 Panasyuk et al., *op. cit.*, in note 34. 499–501.

47 Andreev, *op. cit.*, in note 4. 102, 103.

48 Andreev, *op. cit.*, in note 4. 106; In the end of 1942, the government established a secret educational facility, "the Faculty of Physical Engineering" of MSU, so as to train engineers needed for the developments of the newest weapons (later, focused to the nuclear weapon development) on the basis of the Moscow Mechanics Institute affiliated to the People's Commissariat of Ammunition. Leontovich was in charge of some lessons there with Lev Artsimovich. This "Faculty" could not work well in the severe conditions in the war. However, later, this "Faculty" was separated from MSU and became an independent institute, as the Moscow Physical Engineering Institute (Panasyuk et al., *op. cit.*, in note 34. 499, 501).

49 Panasyuk et al., *op. cit.*, in note 34. 500.

50 Andreev, *op. cit.*, in note 4. 105; Later, Fok wrote a paper titled "On the groundlessness of the research on the generalization of A.A. Vlasov's theory on plasma and solid matter," together with Leontovich, Landou, and Ginzburg. This paper was published in the journal, *ZhETF*, Vol. 16, No. 3, 1946.

51 Ibid., 102–105, 110, 280, 281.

52 Rossiiskii Gosudarstvennyi Arkhiv Sotsial'no-Politicheskoi Istorii (RGASPI), F. 17, Op. 127, D. 275, 1–19.

53 RGASPI, F. 17, Op. 125, D. 275, 79–86.

54 Ibid., 96–106.

2 "Ideology" or "harassment"? "The Physics Conference" unopened

In the previous chapter we discussed the material conditions behind the confrontation between the two groups of Soviet physicists in the early days of the Cold War. Here, we examine the ideological dispute among these physicists which culminated in the unopened Physics Conference planned to be held in 1949.

1 Strife on Khaikin's *Mekhanika* as a preliminary skirmish

The textbook, *Mekhanika* (*Mechanics*), written by Semyon Khaikin (1901–1968), a professor at Moscow State University (Hereafter, MSU), was adapted by the Committee for the Affairs of Higher Education affiliated to the Council of People's Commissars of USSR as a textbook for university students in physics faculties and normal school students in physics and mathematics faculties. It was issued in 1940, with the official seals of the Committee. Nevertheless, in accordance with the intensification of war, it became difficult to continue publishing this textbook. After the return of the universities and other higher education facilities to Moscow and other cities in Europe part with the improvement of the situation, the State Technological Publisher proposed the republication of this textbook in 1944. Khaikin prepared the second version of his textbook for the republication. The new draft was reviewed by some colleagues of the General Physics Department (in charge of education for the students in lower grades) where Khaikin himself worked. Some party members like Professors Vladimir Karchagin (1887–1948), Samson Gvozdover (1907–1969), and Efraim Reikhrudel' (1899–1992) took part in this review together with Yakov Terletskii (1912–1993) from the Theoretical Physics Department. The General Physics Department adopted their "Conclusion" of their discussion.[1] The "Conclusion" said,

> Despite his attempt to illuminate a number of philosophical questions, as well as the basic principles of mechanics and physics in general, the author does not try to introduce the dialectic-materialistic definitions of matter, movement, space and time, and causality, and that he does not try to expound the necessary elements of epistemology of cognition of dialectical materialism.

> This leaves the examination of philosophical problems and a series of the most important problems of mechanics unfinished and allows the possibility to understand these issues in the mode of subjective idealism.[2]

From that point, the situation became more serious and uncontrollable like a rolling snowball. The discussion on Khaikin's draft was scheduled at a Bureau (a leadership section of a basic Party organization) meeting to be held on January 19, 1945. However, it was postponed due to some delay in preparation of a paper by Karchagin. On March 13, in addition to Karchagin, Terletskii and Vasilii Mikryukov (1904–1962) were also put in charge of the preparation of a draft for the resolution of the Party assembly of the Faculty of Physics. Khaikin's explanation was made public at the Party assembly on October 4, and then, a voting on the adoption or rejection of Khaikin's *Mekhanika* and, at the same time, another voting on the propriety of elimination of Khaikin from the Party were both held at the Party assembly on October 24. Twenty-six members supported the rejection of Khaikin's *Mekhanika* against nine oppositions and seven abstentions. However, only ten of them adopted the elimination of Khaikin from the Party, and 22 members opposed the proposal of Bureau and nine members abstained from voting.[3]

Why did the elimination of Khaikin become an issue in the Party organization concurrently with the criticism against his textbook? The criticism against Khaikin's *Mekhanika* was so complicatedly entangled with his application to join the Party. Nikolai Krementsov, a well-known historian of science, suggests that the campaign for recruits to the Party conducted on a large scale during the war brought about a serious change in the composition of the Party members and made the ideological uniformity of the Party cracked due to a sudden increase in the number of new party members.[4]

Even in July 1943, while the wartime recruitment campaign of the Party was actively conducted, there used to be only 13 full members and four candidates (applicants for full membership on probation) in the Faculty of Physics of MSU.[5] In January 1946, the number of those increased to 143 and 47 respectively.[6] Khaikin also applied for a party membership in response to the Party's call in the factory where he was dispatched for technical assistance. His promotion to full membership was examined at the Bureau of the Party organization of his faculty on October 22, 1945. However, meanwhile, he was often harassed in various ways by some old party members. Khaikin became unable to contain himself and sent the Central Committee of the Party a letter dated May 25, 1945, asking for proper measures.[7] For some old party members who were selected as party members through strict screening, the wartime recruit campaign might bring about "popularization" of the Party and newly recruited members might be very often of low qualification in ideological training and, eventually, might be somewhat contemptible. Furthermore, Khaikin was an "idealist" whose "worship to Machism" was pointed out before.

Khinkin's petition for a fair deal was, however, eventually effective after a long consideration at the Central Committee. The memorandum dated January 8, 1946, in joint signature of Georgii Aleksandreov (1908–1961), Chief of

Propaganda and Agitation Section, and Sergei Suvorov (1902–1994), Chief of Section of Science, to Georgii Malen'kov (1901–1988), a Politburo member, concluded that the Commission, consisting of those who

> never and nowhere had any occasion to learn philosophical issues and have never been referred to as major experts in the field of physics, with reckless valor to speak on behalf of dialectic materialism, embarks on the standpoint so harmful to the principles of today's physics, defending their too simplified views on the process of scientific recognition. The error of the Party Committee also consists in that they trusted the complex theoretical problem concerning the evaluation of the book to the vote of the Party assembly, most of which are incompetent members, including students and technical staff. The incorrect direction of the leadership of the Party organization of MSU, is not seen only in the case of Professor Khaikin. In recent years, there has been cultivated among young scientists, students and postgraduates of MSU a mood of arrogant and disparaging attitude toward important scientific centers of the country such as the Academy of Sciences of the USSR, to the scientific activities of old scientists and foreign scientists.[8]

Secretary of the MSU Party Committee, Vasilii Nozdryov (1913–1995) was dismissed from the post of Moscow City Committee member on his mismanagement in this affair. The resolution on Khaikin's *Mechanics* of MSU Party Committee was withdrawn by the upper party organization and Khaikin came back to a normal party life. The MSU Party Committee, which persisted in their position for a while, came to show their wish for a hand-clapping; at a party assembly on December 12, 1946, Vice-secretary of the MSU Party Committee, Sergeev (his name, patronymic, and other biographical details – unknown) reported, "The Party organization of the Faculty of Physics examined Khakin's textbook and Comrade Khaikin revised his draft and presented to the Party committee."[9]

2 The initiation of the postwar "Discussions," "August Session" of VASKhNIL and physics

Mikhail Suslov (1902–1982) became the head of the Propaganda and Agitation Section of the Party's Central Committee, after its chief Georgii Aleksandrov (1908–1961) was criticized during the "Discussion in Philosophy"[10] in 1947 and made to resign that post. Suslov was, however, so busy also with the tasks concerning the Party's international affairs after the sudden illness of a Politburo member, Andrei Zhdanov (1896–1948. After a leave of absence from his duty since July 1, died in August 31). Therefore, Suslov's first deputy, Dmitrii Shepirov (1905–1995), took leadership of the Party's agitation and propaganda activities during a series of "Discussions." Besides him, young Yurii Zhdanov (1919–2006), a son of Andrei and then a son-in-law of Stalin, and a young philosopher Bonifatii Kedrov (1903–1985), then, Editor-in-chief of a newly launched journal, *Voprosy Filosofii* (*Problems of Philosophy*), played central roles in the Party's science policy-making.[11] What was their motivation to initiate such an ideologically charged campaign? Did they have other intentions beside the intention to drive scientists ideologically toward the Cold War? Even archival documents recently declassified do not reveal their real intention. The memoir of Shepirov has no mention about any other orientation of the political leadership than the general ideological unification at the beginning of the Cold War.[12] Therefore, the "Discussion" campaign has been interpreted in various ways, as seen in the Introduction to this book.

Among "Discussions" in various fields of science, the most serious was, no doubt, the case in the biological science. Lysenko doctrine, pseudo-science, was authorized as the sole orthodox theory in "August Session" of All-Union Academy of Agricultural Sciences named after Vladimir I. Lenin (*Vsesoyuznaya Akademiya Sel'skokhozyaistvennykh nauk im. V.I. Lenina*. Hereafter, VASKhNIL) held in 1948, which would obstruct normal development of genetics and other biological sciences for almost a couple of decades. It is now known that the consequence of "August Session" became quite unlike the initial intention of the policy-makers in the Party's science administration by the direct interference by the dictator, Stalin.[13] Stalin, who believed in Lamarckian theory on heritability of acquired characteristics and others, agreed with the insistence of Trofim Lysenko (1898–1976) on the whole, despite denying the argument on "class-oriented essence" of science and excluding excessive ideological rhetoric seen in the draft prepared by Lysenko for the VASKhNIL Session.[14]

The resolutions of "August Session" were also approved by the Academy of Sciences after the so heated discussion in its Presidium meeting.[15] This dispute in Presidium was followed by a chain of disputes on "the situation of biological sciences and their subjects" in meetings of academic councils (decision-making organs in institutes and universities) in various institutes affiliated to the Academy. "Idealism in modern sciences," "isolation from the practice," "kowtow to foreign achievements," and "cosmopolitanism" were heavily criticized in those meetings.

Subsequently, also in the field of physics, a letter dated December 3, 1948, jointly signed by Sergei Kaftanov (1905–1978), Minister of Higher Education, and Sergei Vavilov (1891–1951), President of the Academy of Sciences, appealed to a wide range of physicists to hold a large, all-union-scale conference, the subjects of which would be "struggle against idealism in physics" and "struggle against kowtow to foreign achievements." The conference would be held in March 1949, after the preparation for three months (since December 20, 1948, to March 16, 1949), with many philosophers invited in addition to physicists.[16] Kedrov supposed that such a conference was planned by only a few bureaucrats in Ministry of Higher Education. He also remembers a remark made by Vavilov: He said, "I agreed with them only to eradicate such an impulsive, jesting way."[17]

The start of the "Discussion" campaign, for the first time, in the field of philosophy, reactivated the Fizfak group, which kept silent for a while. At a meeting of the academic council of e Faculty of Physics of MSU as early as in November 13, 1947, Aleksandr Predvoditelev (1891–1973. Then, Dean; Figure 2.1), Vladimir Kessenikh (1903–1970), Nozdryov, and several other professors and teachers launched an attack on some leading then physicists, utilizing "philosophical" discourse. They later called themselves "patriotic and materialistic physicists."[18]

Figure 2.1 Predvoditelev (right) and his students.

Source: Moskovskii Universitet v Velikoi Otechestvennoi Voine (2-e izdanie i dopolnenie). Moskva: MGU, 1985/ 91.

3 The impact of "August Session"

Here, we shall examine the courses of arguments in three institutes related to the field of physics: the Institute of Physics named after Pyotr N. Levedev (1866–1912), in which many physicists who became the targets of their criticism were enrolled; the Institute of Chemical Physics, which had Nikolai Semyonov (1896–1986. A Nobel Prize laureate in Chemistry in 1956) who was heavily criticized as the leader; and the Institute of Physical Chemistry, which was headed by Aleksandr Frumkin (1895–1976) who was also severely criticized.

3.1 Discussion in the Institute of Physics

A philosophical paper[19] which was written by Moisei Markov (1908–1994), a young researcher in the Institute of Physics, on the persistent recommendation of Vavilov and published in the journal *Voprosy Filosofii* (*The Problems of Philosophy*), on his obstinate instigation, was welcomed by many of physicists as a monumental work which reconciled quantum dynamics with Marxism. Although even in the 1930s, Boris Gessen (1893–1936) and others tried to give a Marxist reinterpretation, i.e., justification of quantum dynamics, their trial was suspended by the Great Terror which involved Gessen himself. Also in an early day after the war (November 25, 1946), at an assembly dedicated to ideological problems in MSU, Rector Il'ya Galkin (1898–1990) referred to "the danger to lose the advantageous position of materialist dialectics" for physicists and, at the same time, "the danger to underestimate quantum dynamics, a new and powerful mean to understand nature" for philosophers and appealed to the colleagues to take appropriate steps.[20] It is said that Shepilov realized the necessity to hold "the Physics Conference" when he read the paper written by Markov.[21] That means that he, the initiator of the "Discussion" campaign, assumed more deepened understanding of physicists on dialectic-materialist interpretation of modern physics, which would be obtained through reading the Markov paper and others, as the final target of "Discussion" in the field of physics.

Immediately after the publication of the Markov paper, a conservative philosopher, Aleksandr Maksimov (1891–1976), objected to it in a newspaper, *Literaturnaya Gazeta* (*Literature Newspaper*), in the same tone as before in the 1930s, regarding quantum dynamics as "idealist distortion."[22]

A meeting of the academic council of the Institute of Physics dedicated to the discussion on "the situation of biological sciences and their subjects" was held in November 1, 1948. At the beginning, Director Vavilov stated,

> Comrades! The relationship is very deep and concrete. At first, as I mentioned before, Morganists refer to modern physics and, I have to tell this, many physicists (fortunately they are not Soviets) also take part in the course of the matter.[23]

The following arguments there seem to have been centered around the problems raised in Markov's paper. Vavilov said, "A paper by Moisei Aleksandrovich Markov was issued. This paper includes, no doubt, a serious error besides correct propositions."[24] However, without suggesting the concrete content of this "serious error," he continues,

> The reason why I took up Markov's paper as a topic is not because he committed an error. He made a good deed. He submitted the problems for debates. However, sorry to say, the argument had an entirely incorrect character; especially, our collective of the Institute very little dealt with this paper. Without any doubt, we must make a reference to this paper, not because Comrade Maksimov does not like something in this, but because the most important problem is referred to in this paper. One of the concrete subjects for the forthcoming week or month is to peruse the paper by Moisei Aleksandrovich Markov so earnestly and minutely.[25]

At the platform Dmitrii Blokhintsev (1908–1979) recalled the past:

> The discussion in 1937 was so widely concerned with the problems on methodology of physics. It, no doubt, brought about negative effects in spite of some positive effects. The discussion was conducted in the atmosphere in which many physicists who were interested in the methodological problems kept silent for a long time.[26]

"The discussion in 1937" here means the campaign to hold "the Philosophy Session" in the Academy of Sciences to convict the theory of relativity and quantum dynamics for their "physical idealism" with the Great Terror which began after the Zinov'ev-Kamenev Case in August 1936. If that "Session," for which conservative mechanical engineering scientists Vladimir Mitkevich (1872–1951), Maksimov, and others took leadership, had been held, it might have brought serious damage to the Soviet physics.[27] The inefficient treatment of the matter by the physicists in those days was also rooted in their lack of proficiency in Marxian philosophy. The remark of Blokhnitsev made the audience remember the past only 11 years before and self-reflectively understand the importance of "theoretical fortification" for the defense of modern physics by learning after Markov paper.

After Markov expressed his self-criticism, however, only superficially, saying, "My error consists that I did not pose clearly and definite the problems that should be submitted to the arguments."[28] Subsequently, Kedrov, who was invited to the meeting as the editor of the journal *Voprosy Filosofii* and a researcher in the Institute of Philosophy of the Academy of Sciences, stated,

> Please take the examples of a series of the physicists like . . . (Pascual Jordan – 1902–1980 –) hostile toward us. Why do they blame the supporters of dialectic materialism? (They do so) for that the supporters of dialectic materialism will not accept quantum dynamics and persist on mechanistic materialism. They say that mechanistic materialism was correct in the eighteenth century, but it has already become out of date. Maksimov helps them by criticizing quantum dynamics from the viewpoint of mechanistic materialism.[29]

His remark became the ending remark to that meeting. The "Discussion" in the Institute of Physics, in which the initiator of the "Discussion" campaign himself played the role of coordinator in the discussion, ended in the atmosphere of prearranged harmony.

3.2 Discussion in the Institute of Chemical Physics

The Institute of Chemical Physics was established in 1940 as one of the affiliated organizations of the Academy of Sciences on the basis of its forerunner which had been affiliated to the People's Commissariat of Petroleum Industry. Nikolai Semyonov was Director continuously since 1931 when the forerunner still existed. He achieved outstanding results in the field of chemical kinematics of gas phase materials, especially chain reaction mechanism in explosion and detonation.[30] Therefore, in his Institute the research concerning explosives and explosion mechanism was intensively carried out during the war. Lev Landau (1908–1968. A Nobel Prize laureate in Physics in 1962) was also dispatched to this Institute from his own Institute of Physical Problems and gave a presentation titled "To the Theory of Ignition of Solid Explosive."[31] Thanks to the scientific contribution to the war efforts, the scale of the Institute was almost doubled during the war. In 1944–1945, the Institute included totally four departments and 14 laboratories: four laboratories in the department of chemical kinematics, three laboratories in the department of explosives, five laboratories in the department of combustion in internal combustion engines, a laboratory in the department of combustion, and another laboratory.[32] As well known, after the war, Yurii Khariton (1904–1996), Yakov Zel'dovich (1914–1987), Mikhail Sadovskii (1904–1994), Aleksandr Ratner (1906–1956), and others from this Institute played crucial roles in the Soviet nuclear weapon development.

In this Institute, the discussion on "the situation of biological sciences and their subjects" was conducted at the enlarged meeting of the academic council held in October 29 where totally 74 researchers attended. According to the summarized record of the meeting, in his key-note speech, Deputy Director Viktor Kondrat'ev (1902–1979) pointed out some defects about which the Institute had to reconsider itself self-reflectively: unsatisfactory situation on relationship with practice, exclusiveness in some thoughts, and the lack of vivid examination of the activities of the Institute. After that, Director Semyonov pointed out the ideological difference between the bourgeois science and the Soviet science, stating that the bourgeois theoretical chemistry could not make any significant achievement for the quarter of a century.[33]

In the discussion, A.I. Sokolik, a chief of a laboratory and a doctor in chemical sciences, was called upon to make a remark and expressed his disagreement with the basic line of the key-note speech. According to the summarized record, he "disagreed with the conclusion on the practical direction of the group of researchers in chemical kinetics (which would lead to – *Ichikawa*) weakening the front-line theoretical activities, which might activate formalist course (N.S. Akulov)."[34] In this case, "formalism" is a jargon which means the insistence of Nikolai Akulov (1900–1970) who persistently accused Semyonov.[35] However, this remark of Sokolik sounded like disagreement with the basic science policy of the Soviet Union warning isolation of scientific research from practice in the term of "formalism."

What is more interesting is that his remark was supported by the majority of the attendees. In addition to a commonplace remark of "self-reflection" of the whole Institute, the resolution of the meeting points out "the necessity to overcome" "formalism," saying, "the Institute had been reconciled to the situation in which formalist trend, developed by a professor at Moscow State University, Akulov, in the research field of chemical kinetics, still exists in various fields of the Soviet chemistry."[36] Also in the enlarged meeting of the academic council held subsequently in November 2, counter-criticism against the criticism launched by Akulov was taken up as a major topic. Zel'dovich uttered, "It is needed to put an end to the struggle with the distortion of chemical kinetics by Professor Akulov."[37]

The resolute attitude this Institute showed seems to have been rooted in self-confidence obtained through the contribution to the war efforts of the state.

3.3 *Discussion in the Institute of Physical Chemistry*

Colloid Electrochemical Institute was established in Moscow in 1934 on the basis of its forerunner, Colloid Electrochemical Laboratory established in Leningrad in 1930. The institute was reorganized as the Institute of Physical Chemistry in October 12, 1945, based on the decision of the Presidium of the Academy of Sciences dated March 29 of that year.[38] In 1939, Frumkin took the responsibility as Director. He studied in Strasburg and then in Wisconsin University in USA in 1928. During the war he was engaged in the development of electromagnetic method of uranium isotope separation independently from Academy's "Second Laboratory" which played the crucial role in the early days of the Soviet nuclear development project. After the war (since 1946), he worked for the nuclear development in "the Special Laboratory of Radiation Chemistry" established in secret within his institute.[39] At the moment of reorganization as the Institute of Physical Chemistry, it had four departments (electrochemistry, heterogeneous catalyst effects and topochemistry, dispersions and colloids, corrosion) and five laboratories (adsorption process, photochemistry, interfacial forces, electronic imaging and X-ray photography, and Ex-director Kistyakovskii's own laboratory) and included two full members of the Academy of Sciences, three corresponding members, 14 doctors, and 35 doctoral candidates.[40]

What was striking about this institute was the presence of an opponent group against the institute's leadership. For example, P.I. Dankov, the chief of the laboratory of electron imaging and X-ray photography, often sent letters complaining about Frumkin to the Party's Central Committee.[41]

In the case of this institute, a meeting of the academic council dedicated to discussion on "the situation of biological sciences and their subjects" was held on October 2, 1948.[42] For every attendee was obliged to read his or her own notes previously prepared for this meeting, it was so prolonged. In the discussion, the evaluation of Semyonov's theory on chain chemical reaction mechanism was argued (Frumkin supported it as "a materialist achievement"). Suddenly, a young colleague, Dolin (his given name and patronymic – unknown) uttered, "For these three years, while I have worked in the institute, I have often heard Professors Dankov, (I.V.) Krotov and (Kseniya) Gorbunova criticize Frumkin's scientific works. Their criticism was originated from various aspects. In my opinion, they are neither conscientious nor principled."[43] Dolin's remark, however, rather stimulated the trend of criticism against Frumkin. Dankov pointed out Frumkin's "kowtow to foreign science," citing Frumkin's "silence on the priorities of (Mikahil) Lomonosov and (Dmitrii) Mendeleev" and overestimation of the achievement of Irving Langmuir (1881–1957. A Nobel Laureate in Chemistry in 1932) as the evidences for his argument.[44] Frumkin objected to Dankov, saying,

> It is no use to cite from my paper. I referred to Langmuir as a courtesy. This courtesy was demanded by the historical stage when it was written. This paper was not written today. Immediately after the end of the war, it was written. During the war, Langmuir kept the pro-Soviet position apparently. When he visited Moscow, he made a speech properly close to our view over the radio. As for his contribution to science, there is no doubt he is a great scientist.[45]

During World War II, the United States was in alliance with the Soviet Union. Then the Soviet scientists ardently desired to exchange with American scientists. Such an explanation by Frumkin reflected directly the true feelings common with many Soviet scientists.

In this case, however, denouncing Frumkin for his "worship of foreign science" was somewhat effective. In his report to the Organization Bureau of the Party's Central Committee, Yurii Zhdanov wrote,

> Propaganda and Agitation Section is calling the attention of the Presidium of the Academy of Sciences to the serious defects in the activities of the Institute of Physical Chemistry. Following the direction of our division, now reexamination on the editorial board of the journal, *Fizicheskaya Khimiya* (*Physical Chemistry*), and change of editorial staffs are being carried out. Scientists acting for various fields of chemistry are included as the editorial staffs. The presidium of the Academy of Sciences removed Dubovitskii from his post as Deputy Director of the Institute of Physical Chemistry and appointed Viktor Spitsyn, a chief of a department in MSU, to that post. A full member of the Academy, Frumkin was summoned to Agitation and Propaganda Section and was pointed out the defects in the activities of the Institute of Physical Chemistry. He admitted that he underestimated the classics in Russian chemistry. Propaganda and Agitation Section proposed to conduct more detailed investigations on the activities of the Institute of Physical Chemistry and Institute of Chemical Physics and to inform a secretary of the Party's Central Committee of this matter.[46]

Consequently Frumkin was dismissed from the post of Director.[47]

Meetings of the academic council of the Institute of Physical Chemistry were held totally 20 times in 1949. Four of them were dedicated to discussion on "the history of development of physical chemistry and the criticism to the scientific views made by foreign bourgeois physical chemists."[48]

4 "The Physics Conference" unopened

4.1 Preparation for "the Conference"

While the discussions were carried out in various institutes in the Academy of Sciences, the preparation for "All-Union Physics Conference" was making rapid progress. Immediately after the first call for the conference in the letter jointly signed by Kaftanov and Vavilov, the arrangements committee was organized with Aleksandr Topchiev (1907–1962), close to Shepiliv, as Chair. In the first meeting of the committee, Kaftanov read his speech on the outline of the conference, full of intimidation upon the modern physicists by reconfirming the philosophical position stated in Lenin's *Materialism and Empirio Criticism*, criticizing idealism and cosmopolitanism, referring to the possibility to "make fame of the theories lose their prestige from the standpoint of dialectic materialism" and proposing the radical reexamination of the way to train successors and personnel.[49]

According to a prominent historian of science, Vladimir Vizgin, in the meetings of this committee held so frequently, those physicists who engaged in the forefront of modern physics such as Yakov Frenkel' (1894–1952), Vladimir Fok (1898–1974), Tamm, Landau, Khaikin, Vitalii Ginzburg (1916–2009. A Nobel Prize laureate in Physics in 2003), Markov, Leontovich, and others were criticized on idealism and cosmopolitanism by "a philosophical conservative," Maksimov, Arkazhii Timiryazev (1880–1955), a mechanist-materialist who played an active part in the philosophical debates in 1920s, and other philosophers and "patriotic and materialistic physicists." They even adopted an analogy from biology for their criticism against the theory of relativity and quantum dynamics, labeling them Morganism and Weissmaninsm. Tamm, Ginzburg, and other criticized physicists made great efforts to refute, insisting that each of the theories of relativity and quantum dynamics "reflects the objective world"; however, they were not so successful. The main issue of the argument shifted from the criticism of physical idealism at the early stage immediately to the criticism against cosmopolitanism and kowtow to foreign science. Even Vavilov's conciliatory remark that the fundamental significance of the progress of physics consists in the technological progress met the hard criticism repeatedly made by totally eight "Fizfak" physicists and philosophers. The meetings of the arrangements committee appeared as if they had been kangaroo courts.[50] However, according to Ethan Pollock, a Kaftanov speech at a preparatory meeting on January 15, 1948, in which he compared Albert Einstein (1879–1955) with Lev Tolstoi (1828–1910), not August Weismann (1834–1914) or Thomas Morgan (1866–1945), changed the tide in the discussion.[51] Markov, Fok, Landsberg, Aleksandr Andronov (1901–1952), Tamm, and other physicists in the side of the Academy succeeded in providing effective counterarguments. Nozdryov made his speech in the meeting on March 7 which became the last remark from the side of "patriotic and materialistic physicists." Kaftanov, who took the platform after Nozdryov, disappointed "patriotic and materialistic physicists" by pointing out "the errors" of Maksimov and Timiryazev and concluding suddenly and insubstantially that all had to "make efforts for the progress of the Soviet physics."[52]

Although at the meeting of March 16, Kaftanov repeated that the full session of "Conference" would be held on the 21st, on the day, he himself asked a vise-primer, Malen'kov (in charge of culture and science in the Party's Central Committee after the death of Andrei Zhdanov), to postpone the Conference until May10. However, eventually the full session of All-Union Physics Conference was not held even on that day of May 10. A letter dated April 9 of Shepilov, the acting chief of the Propaganda and Agitation Section of the Party's Central Committee, entreated Malen'kov to abandon to hold the Conference, because "the preparation for the Conference is insufficient and the deeper examination of the problems and the concrete proposals are urgently needed." Indeed, accepting the urging of Vavilov, Shepilov himself made up his mind not to open such a conference.[53]

4.2 Sergei Vavilov as a key person

During the war, at almost the same time with the "erroneous homicide" upon his elder brother, a prominent geneticist, Nikolai,[54] Sergei Vavilov was appointed as an authorized specialist (*upolnomochennyi*) for the State Defense Committee, i.e., the wartime temporary supreme state organ of the Soviet Union.[55] As soon as he was appointed, he submitted a report to Stalin on the significance of nuclear research and cosmic ray research and the situation of the research in the USA, Great Britain, France, and Germany.[56] Since that moment, he took a great deal of pains so as to establish a good relationship with the political power. After the war, on the initiative of Stalin, he replaced Vladimir Komarov (1869–1945), who intimated his intention of resignation from his post of President of the Academy of Sciences on July 17, 1945.[57] Under Vavilov's leadership (Figure 2.2), the Academy of Sciences turned from a somewhat detached, autonomous organization indifferent to occasional subjects of the state into a practical organization making contributions to the state scientific and technological projects from the

Figure 2.2 Sergei Vavilov at a Presidium meeting of the Soviet Academy of Sciences.

Source: Courtesy of the Archive of Russian Academy of Science.

viewpoints of specialists by establishing within its system various ad-hoc organizations such as "the Special Academic Council" (which played an important role in nuclear research and the developments of some nuclear reactors), "the Special Commission on Cosmic Problems," commissions on the constructions of hydro-electric power stations in Samara (then, Kybyshev) and Volgograd (then, Stalingrad), on canal building in Turkmenistan and others. Simultaneously with these steps, the Academy was enlarged in its scale by establishing the Academies for each republic in the Soviet Union, seven branches in the Russian Republic, and other organizational expansions. What is the most thought-provoking here is the establishment of the Academic Secretariat affiliated to the Presidium on March 17, 1949, which was a standing organ inspecting the practices of the plans in various fields of scientific research.[58]

Already on February 26, the plan to arrange the personnel for the Secretariat was approved in the Party's Central Committee. Aleksandr Topchiev was to be appointed as Chief Academic Secretary, and Yurii Zhdanov (1919–2006) was assumed to be a member of the Secretariat.[59] As today's eminent historian, Vladimir Esakov here supposes the existence of a sort of "secret treaties" between the Party (maybe, Shepilov) and Vavilov.[60]

5 The outcome of "the Discussion" in the field of physics

As we examined in the previous chapter, many of the professors and teachers at the Faculty of Physics had very difficult lives during the war, having engaged in the evacuations initially to a very far place and then to another place, in the wartime research not so effective with very poor research means and in the education of a great deal of students overflowing the narrow campus in a very severe condition. On the other hand, their colleagues who worked only part time in MSU were evacuated to other places and engaged energetically and effectively in the wartime research in relatively rich conditions there. Thus, the conflicts between two groups of the Soviet physicists grew large during the war. The antagonism accumulated in the side of full-time teachers of MSU against the part-time teachers mobilized to the wartime research in another place was looking for "an outlet."

The "Discussion" campaign in various fields of science provided an outlet. The main theme of the "Discussion" must have been a struggle against philosophical idealism or overcoming the bourgeois science at least in the initial stage of the "Discussion" in the field of physics. However, as seen in the examples we examined here, what pressured the physicists the most during the "Discussion" was neither the struggle against idealism nor that against the isolation of scientific research from the practice, but the campaign of anti-cosmopolitanism or the criticism against kowtow to foreign science. In addition, the arguments in the letters sent to the Party's Central Committee in 1948 to accuse some physicists of their "harmful" roles were also concerning the criticism against cosmopolitanism and the worship to foreign science which fawned upon the Western science and ignored the achievements of the Russian science.[61] Vizgin regards that the naïve argument regarding quantum dynamics as "physical idealism" became invalid, because the paper written by Markov succeeded to a certain extent in interpretation of quantum dynamics from the viewpoint of dialectical materialism.[62] In addition, we must take into consideration the timing of the "Discussion." Different with the prewar time, quantum dynamics was already accepted by a wide range of physicists including Fizfak physicists. For example, Sergei Vasil'ev (1904–1999), who was a specialist in particle acceleration but supported the "patriotic and materialistic physicists," was accused of preaching in his lectures "idealist views and theories" after Niels Bohr (1885–1962. A Nobel Prize laureate in Physics in 1922), Ernest Rutherford (1871–1937. A Nobel Prize laureate in Chemistry in 1908), Linus Pauling (1901–1994. A Nobel Prize laureate in Chemistry in 1954, Peace Prize in 1962), and Erwin Schrödinger (1887–1961. A Nobel Prize laureate in Physics in 1933).[63] In addition to their own poor knowledge on philosophy or inexperience in philosophical thinking,[64] the "patriotic and materialistic physicists" could not obtain necessary assistance concerning philosophical aspects from the side of philosophers except for an old-fashioned philosopher, Maksimov.

Nevertheless, what is the most important there is that the Party's policymakers on science policy approved of the criticism from the "patriotic and materialistic physicists" against cosmopolitanism and kowtow to foreign science, as seen typically in the argument in the Institute of Physical Chemistry.[65] It seems that, careful not to harm scientific research itself,[66] the Party's administrators in charge of scientific matters drove the scientists into "chorus" for the struggle against cosmopolitanism and worship of foreign science, making use of the "patriotic and materialistic physicists" as tricksters.[67] The break-out of the Cold War means the change in position of the United States for the Soviet Union from its alliance to animosity. During the war while the United States was an ally of the Soviet Union, the Soviet scientists such as Frumkin could engage in scientific exchange with American scientists and enjoyed the benefits of progressive American science. Many Soviet scientists expected much of scientific exchange with American scientists. In April 1945, a plan to dispatch to the United States Artyom Alikhanyan (1908–1978) and Vladimir Veksler (1907–1966) for the study on cyclotrons and Lev Artsimovich (1909–1973) for the study on electronics (for electron microscopes, infrared apparatuses, and others) was settled on and was approved by the Propaganda and Agitation Section (eventually denied at the Politburo). Also in 1945, the American scientists offered to hold a scientific conference in New York with 25–75 scientists invited from the Soviet Union. Even after the war, in 1947, Vavilov planned to hold "scientific conference on theoretical physics" in Leningrad, inviting many prominent foreign physicists.[68] No archival document reveals the real intentions of the Party's administrators in charge of scientific matters. However, it may be guessed that they supposed further progress of scientific exchange with the American scientists might lead the Soviet science to "the lee side" to the American science.

If the purpose of the "Discussion" campaign had consisted in the acquisition of a wide-range consent among scientists to the superiority of the Party's authority over the autonomous authority of the scientists themselves or, more generally, the ideological subsumption of scientists, as Yurii Krivonosov, a prominent historian of science, considered,[69] the Party's administrators would not need to hold the full sessions of "All-Union Physics Conference," even taking a risk to "throw the baby with bathwater." It can be said that Shepilov attained his goal with only the preparatory meetings and succeeded in sending some Party apparatchiks to the administration of the Academy of Sciences.

On the other hand, the physicists involved in the "Discussion" experienced a serious and considerable change. Vladimir Fok, a leading physicist in Leningrad, is known as a courageous figure who sent his letters requesting to take proper measures to protect modern physics to the Party' Central Committee, the Presidium of the Academy of Sciences, and others even in the turbulence of "the Great Terror."[70] He did not attach any political or philosophical remarks to his commentary he wrote for a newspaper under the title, "*Problemy atomnogi fiziki* (the Problems of Nuclear Physics)" around 1936.[71] However, when "the Physics Conference" was being prepared in 1949, he issued a paper titled "Osnovye zakony fi ziki v svete dialekticheskogo materializma (Fundamental laws of physics in Light of Dialectical Materialism)" in *Vestnik Leningradskogo Universiteta.* (*Bulletin of Leningrad State University.*) No. 9, 1949. The document file on him which is kept in Sankt-Peterburg Branch of the Archive of Russian Academy of Sciences includes several handwritten drafts and typed manuscripts for this paper. He tried to rewrite his paper at least four times till accomplishment.[72] In opposition to those who try to give an idealist interpretation to modern physics, he wrote in this paper, "We regard that the theories of physics reflect the movements based upon the laws absolutely existing in nature. They agree with experimental inspections. We regard the margins of errors in experimental inspections tolerable in principle." He also made clear his standpoint putting the greatest significance upon objective reality of the movements of materials by citing Lenin's remark in the book *Materialism and Empirio Criticism*, criticizing Machism which insisted on extinction of materials.[73] His view here well echoed the standpoint of the paper by Markov. Another leading physicist in the Soviet Union in those days, Abram Ioffe began to write his draft for his address which was scheduled to be read at a meeting of his institute, the Leningrad Institute of Physical Technology, which would be held on March 25, immediately after the postponement of "All-Union Physics Conference," with a warning against the stratagem devised to harm the prestige of materialist philosophy and the Soviet science and to spread wide the American way of cosmopolitanism under such a situation in which an ideological attack was being developed as the preparation for the U.S. invasion of the Soviet Union.[74] The Soviet physicists were objectively demanded to pledge their ideological allegiance to the Party by speaking "Newspeak," the communist jargon, and in such an atmosphere the "Discussion" campaign fermented.

6 Destalinization and the Soviet scientists

In October 1953, under the very unique atmosphere brought about by the death of Stalin and the arrest of Lavrentii Beriya (1899–1953), at a customary meeting of the representatives of the classrooms of Komsomol organization in MSU which gathered almost 400 students, there suddenly arose among the participants an unexpected intense criticism against several "substandard teachers." The news of this incident was brought to the Party's Central Committee. The necessity to investigate this incident from the viewpoint of the Party was appealed by a letter dated November 12, jointly signed by P. Ponomarenko (Minister of Culture), Vyacheslav Maryshev (1902–1957, Minister of Middle-scale Machine-building, i.e., Minister in charge of nuclear industry), Aleksandr Nesmeyanov (1899–1980, President of the Academy of Sciences in 1951–1961), and Mstislav Keldysh (1911–1978, later, President of the Academy), and addressed to the Politburo members, Georgii Malen'kov, and Nikita Khrushchev (1894–1971). According to the conclusion of the commission established so as to investigate the activities of MSU, so-called Malyshev commission, the government decided to discharge Arsenii Sokolov (1910–1986) from his post of Dean of the Faculty of Physics (Vasilii Fursov – 1910–1998 – succeeded him) and partly change the personnel of MSU.[75] Due to this means, the "patriotic and materialistic physicists" in the Faculty of Physics lost their influence.[76]

Here, we shall take a view on the post-Stalin turbulence in wider range of the Soviet scientists.

An outcome of the "Discussion" campaign was the enhancement of the role of those philosophers and party ideologues who tried to manage scientists and intellectuals with the help of "Newspeak." As an ideological "watchdog," they had been frightening to the entire intelligentsia. One of the results of their "self-perpetuating" expansion of the Cold War propaganda discourse, the philosophers and ideologues gave their attention to the "bourgeois and reactionary essence" of modern Western thoughts and took it their main task to annotate and criticize such ideas.[77] In 1951, the Institute of Philosophy of the Academy of Sciences issued a comprehensive work on that problem. In this collective work titled *Protiv Filosostvuyushchikh Oruzhenostsev Amerikano-Angliiskoi Imperializma: Ocherki Kritiki Sovremennoi Bruzhuaznoi Filosofii i Sotsologii* (*Against the Philosophizing Followers of American-British Imperialism: Essays on Criticism of Contemporary American-British Bourgeois Philosophy and Sociology*), they analyzed several ideological currents such as phenomenology, pragmatism, and so on.[78] Furthermore, in 1953, the Institute of Philosophy (its Section for the Criticism against Contemporary Bourgeois Philosophy and Sociology) listed totally 17 targets of their criticism, such as Cybernetics, Neo-Kantianism, and others.[79] In the midst of such uplift of ideological activities by philosophers, Stalin suddenly died.

Only eight days after the death of Iosif Stalin, a meeting of the Presidium of the Academy of Sciences issued a decision to postpone readily the compilation of a voluminous collective work, titled *Dialekticheskii Materializm i Sovremennoe Estestvoznanie* (*Dialectical Materialism and Contemporary Natural Science*), the publication of which had been scheduled already in November 21, 1952. The reasons for this decision were given as there as following; (1) The views of the members of the Academy were not in accord on various problems and more beforehand examination was needed. (2) The members who were to be in charge of the editorial work of the book were so busy with their main occupations. (3) A significant portion of authors, especially those who belonged to the Department of Chemical Sciences and the Department of Biological Sciences, could not set about the task of writing their papers due to their heavy burdens. In the place of this collective work, the compilation and publication of a book, titled *Klassiki Marksisticheskoi Teorii ob Estestvoznanii* (*Classics on Marxist Theory on Natural Science*), was entrusted to the Institute of Philosophy. The Presidium obliged its Departments of Physical and Mathematical Sciences, Chemical Sciences, Biological Sciences, Geologic and Geographic Sciences, and Technological Sciences to publish papers and books concerning philosophy on natural science continuously also in 1953–1954. Despite that, the Presidium decided, at the same time, to begin the compilation of a new book under the title *Dialekticheskii Materializm i Sovremennoe Estestvoznanie* (*Dialectical Materialism and Contemporary Natural Science*) in 1954[80]; however, this subject was not discussed any more at least until the end of 1956.

Approximately one month after the death of Stalin, an eminent painter, Igor' Grabar' (1875–1960), a distinguished geologist-paleontologist and popular writer, Vladimir Obruchev (1863–1956), and a still young but already recognized mathematician, Sergei Sobolev (1908–1989; the last two persons, they say, spoke on behalf of "the scientists working in the Academy of Sciences and the Moscow State University"), sent a letter dated April 6, 1953, with their joint signatures to Secretary of the Party's Central Committee, Nikita Khrushchev; in the letter they complained about the behavior harmful to natural science of Yurii Zhdanov, then Chief of the Central Committee's Division of Natural Science, Technology and Higher Education Facilities, and demanded a careful survey of his behavior. Such a courageous petition of the three persons might be effective; Yurii lost his post as Chief at the opportunity of reorganization of the central departments in which his Division was unified with another division into a single division, Division of Science and Culture.[81]

Dismissal of Georgii Aleksandrov (the former Director of the Institute of Philosophy) from the post of Minister of Culture USSR in 1955[82] led to outbreaks of antagonistic emotion of the scientists to the philosophers. At a meeting held June 3, 1955, the Presidium of the Academy of Sciences sharply criticized the activities of the Institute of Philosophy for delays in its work, negligent researchers who didn't write any articles for more than one year, and lack of governability of the former Director, Aleksandrov. Beside them, the Presidium severely criticized the Institute for its harmful role to natural science, saying,

> The lessons of Lenin and Stalin on the comprehensive study of materialistic dialectics as a philosophical science, the need to strengthen the union of philosophers and natural scientists, the inseparable linkage between philosophical work and vital needs of actual life have not yet become the guiding principles for the activities of the Institute. . . . Some workers of the Institute of Philosophy occupied an incorrect position on the philosophical problems of contemporary natural science. . . . For example, a corresponding member of the Academy of Sciences, Maksimov A.A. (who led the Sector of Philosophy of Natural Science in the past) nihilistically approached the theory of relativity, discarding its valuable physical conclusions, as if it had allegedly

> contradicted the dialectical materialism. . . . Many researchers regarded as their task only to mechanically stick the etiquette 'dialectical materialism' on the statements of individual figures of Soviet science, without critical analysis of their work on the matter.[83]

In the discussion in that meeting, a philosopher, but then out of the mainstream faction, Bonifatii Kedrov, who had been dismissed from the post of the editor-in-chief of *Voprosy Filosofii* and put away from the Institute of Philosophy, critically analyzed in detail the activities of the Institute.[84] Vladimir Fok accused the philosophers for their obstructive role, looking back bitterly to the past: "The attempts to publish the works explaining the problems of natural science in philosophical context have been beset by obstacles that we were able to overcome only with the intervention from the side of the Party's Central Committee."[85] Finally Sergei Rubinstein, a psychologist who was dismissed from the post of a department head in MSU and, at the same time, a section chief in the Institute of Philosophy for his "cosmopolitanism," made an actually ending remark full of critical comments against the Institute of Philosophy.[86]

In his article titled "Za pervoe mesto v mirovoi nauke (Toward the First Position in the World Science)," published in 1956, Nesmeyanov, President of the Academy of Sciences, introduces a tale of "a great physicist," saying,

> (He) seriously argued that the movement of a person passing by a blooming flower changes the speed of growth of a flower, though within a billionth of a second. And none of the members of the editorial board of the journal *Voprosy Filosofii* would explain to the scientist that he preaches idealistic nonsense without no difference with the shaman's creed that his jumps and dances around the sick man influence the course of the disease.[87]

The days of excessive ideological jargon running wild was rapidly passing away.

A large-scale conference named Vsesoyuznoe Soveshchanie po Filosofskim Voprosam Estestvoznaniya (All-Union Conference on Philosophical Problems of Natural Science) was carried out in October 1958.[88] In the end of the conference, each presenter was requested to give his or her "presenter's remark at closing the conference." In his "remark" which would be often cited as a landmark statement, Sobolev declared on behalf of all the Soviet scientists, their refusal of any interference from the side of philosophers, saying, "We cannot divide physics into materialistic and idealistic physics. We cannot say that this atomic bomb is materialistic, but this one is idealistic, that this accelerator of elementary particles is idealistic, and this one is materialistic."[89]

Notes

1 A.S. Ilyushin, *Sud'ba knigi i cheloveka* (unpublished draft), 45–47. The author cites from this draft with the permission of Professor Aleksandr Ilyushin (1943–). However, the commentary titled "Introduction" to Khaikin's letter dated May 25, 1945, to the Party's Central Committee reprinted in the periodical of the Vavilov Institute for the History of Natural Science and Technology of Russian Academy

of Sciences, *Issledovaniya, po Istorii Fi ziki i Tekhaniki*. 2009–2010. (214–224) includes some similar contents that can be seen in this draft by Professor Ilyushin.

2 Akhiv Muzeya istorii fizicheskogo fakuliteta MGU. Delo S.E. Khaikina.

3 Tsentral'nyi arkhiv obshchectvenno-politicheckoi istorii Moskvy (Hereafter, TsAOPIM), Fond (F.) 478, Opis' (Op.) 1, Delo (D.) 92, 4, 6, 8-b, 50-b, 53; Ilyushin, *op. cit.*, 45–47.

4 N. Krementsov, *Stalinist Science*. Princeton and Oxford, Princeton University Press, 1997, 96–99.

5 TsAOPIM, F. 478, Op. 1, D. 57, 1.

6 TsAOPIM, F. 478, Op. 1, D. 114, 123.

7 Delo Khaikina, *op. cit.*, in note 2. For example, a party member, Professor Sergei Mraveiskii, told Khaikin that they would be needed to "prepare for a prayer to mourn," if Khakin did not change his "idealistic" thought. Nozdryov, Secretary of the Party Committee in MSU, blamed Khaikin, saying that he made use of a factory where he was dispatched to as "a bypath," with his "idealism" unchanged.

8 Rossiiskii gosudarstvennyi arkhiv sotsal'no-politicheskoi istorii (RGASPI), F. 17, Op. 125, 363, 72–74.

9 TsAOPIM, F. 478, Op. 1, D. 114, 163.

10 The minutes of this strange event was expounded upon in details in a paper of Yurii Krivonosov (Yu.I. Krivonosov, "Srazhenie na filosofskom fronte: filosofskaya diskussiya 1947 goda – prolog ideologicheskogo pogrom nauki," *Voprosy Istorii Estestvoznaniya i Tekhniki*, No. 3, 1997.

11 A.B. Kojevnikov, *Stalin's Great Science: The Times and Adventures of Soviet Physicists*. London, Imperial College Press, 2004, 204.

12 See, D. Shepilov, *Neprimknuvshii*. Moscow, Vagrius-priroda, 2001.

13 The Academy of Sciences had already decided in May 30, 1945, to establish a new institute named, the Institute of Cytogenetics, beside the Institute of Genetics whose Director was Lysenko, in order to restrict the influence of Lysenko upon biological sciences. Assemblies for critical examination on Lysenko doctrine were subsequently held in MSU in November 1947 and in the Deaprtment of Biological Sciences of the Academy in February 1948 (RGASPI, F. 82, Op. 2, D. 930, 145). However, as soon as Lysenko succeeded in getting Stalin to support him, the situation was reversed. Lysenko sent a petitionary letter, saying, "If the opinion of Yurii Zhdanov was the Party's opinion, please, propose my dismissal from the post of President of VASKhNIL." Stalin stated his unpleasantness in a Politburo meeting in May (A.B. Kozhevnikov, "Igry stalinskoi demokratii i ideologicheskie diskusii v sovetskoi nauke: 1947–1952gg." *Voprosy Istorii Estestvoznaniya i Tekhniki*, No. 4, 1997, 43). According to the memoirs of Shepilov, immediately after he charged Yurii Zhdanov with the tasks concerning the criticism against Lysenko in April 10, 1948, he was summoned to a meeting of the Politburo together with Suslov and Yurii. There, Stalin furiously ordered revocation of the decisions to conduct a campaign to criticize Lysenkoism and entrust the problems to the examination by scientists and specialists (Shepilov, *op. cit.*, in note 12. 128–131).

14 See, Kirill Rossiianov, "Reshaping Political Discourse in Soviet Science: Stalin as Lysenko's Editor." *Russian History*, Vol. 21, 1994, 49–63.

15 Arkhiv Rossiskoi Akademii nauk (ARAN), F. 2, Op. 3, D. 92, 1–152. Kedrov remembers that Sergei Vavilov was much agitated by such a result in spite of his position as President (B.M. Kedrov, "Vospominaniya ob akademike S.I. Vavilove." in V.A. Lektorskii (Ed.), *Bonifatii Mikhilovich Kedrov: Ocherki vospominaniya i materialy*. Moscow, Nauka, 2005, 334: This memoirs was written in 1983–1984, but left unpublished until 2005).

16 V.P. Vizgin, "Yadernyi shchit v 'trinadtsatiletnei voine' fizikov s nevezhestvennoi kritikoi sovremennykh fizicheskikh teorii." *Uspekhi Fizicheskikh Nauk*, Vol. 169, 1999, 1373.
17 Kedrov, *op. cit.*, in note 15. 335.
18 Vizgin, *op. cit.*, in note 16. 1372.
19 M.A. Markov, "O prirode fizicheskogo znaniya." *Voprosy Filosofii*, No. 2, 1948, 140–176: Kedrov, who knew Vavilov in person since his university days, showed a draft of a paper titled "A minstrel of physical idealism" and sent to the editorial boad of his journal, *Voprosy Filosofii*, which tries to criticize quantum dynamics. As soon as Vavilov read this draft, he came to urge Markov who hesitated about the publication of his paper written on Vavilov's advice to hurry to publish it. Markov hoped Maksimov, a conservative philosopher familiar to natural science, would support him, as was often the case with the Soviet physicists in those days. Vavilov finally succeeded in obtaining Markov's agreement, telling him a lie, as if Maksimov had supported his paper ardently (Kedrov, *op. cit.*, in note 15. 332). Today, we can read this paper in the first volume of Markov's Selected Works (M.A. Markov, *Izbrannye Trudy*, V 2-kh tt, Vol. 1. Moscow, Nauka, 2000, 365–407). In this paper, Markov emphasizes on objectivity of the subjects of quantum dynamics which can be recognized through the experimental equipment that translates the "language" of microscopic world of nature to the "language" of macroscopic world of physics.
20 Tsenral'nyi munitsipal'nyi arkhiv Moskvy (TsMAM), F. 1609, Op. 2, D. 189, 33.
21 V.D. Esakov, "K istorii filosovskoi diskussii 1947 goda." *Voprosy Filospfii*, No. 2, 1993, 85, 86.
22 In his paper, "Ob odnom filosofskom Kentavre (On One Philosophical Centaur)," published in the *Literaturnaya Gazeta* (*Literature Newspaper*) of April 10, 1948, Maksimov asserts on formalist logic that the microscopic world itself cannot be recognized objectively, because the physical conditions of particles are influenced by macroscopic equipment. He could not gain the approval of any physicists, however, except for Terletskii (A.S. Sonin, *"Fizicheskii Idealizm": Istoria Odnoi Ideologicheskoi Kampanii.* Moscow, Fiziko-Matmaticheskaya Literatura, 1994, 90–93).
23 ARAN, F. 532, Op. 1, D. 145–631, 1, 61.
24 Ibid., 65.
25 Ibid., 66.
26 Ibid., 74.
27 Vizgin, "Yadernyi shchit . . ." *op. cit.*, in note 16. 1372.
28 ARAN, F. 532, Op. 1, D. 145–631, 1, 90.
29 Ibid., 94.
30 See, A.F. Andreev (Ed.), *Kapitsa, Tamm, Semyonov: Biografii v ocherkakh i pis'makh.* Moscow, Vagrius-priroda, 1998, 408–575.
31 ARAN, F. 342, Op. 1, D. 45–221, 11.
32 ARAN, F. 342, Op. 1, D. 53–241, 8–10.
33 ARAN, F. 342, Op. 1, D. 75–63, 1, 1, 1.
34 Ibid., 2: Sokolik furiously shouted, "I, clearly, disagree with the conclusion of the leadership of the institute, i.e. the conclusion to change the fundamental research on kinetics to the industry-oriented research. . . . Semyonov brought up a large group of researchers on kinetics. Do you tell (us) to turn its direction to another? Do you say that the theoretical research shall be conducted by Akulov and his colleagues?" Soon after, an "expert on philosophical issues" of this institute, S.N. Khudyakov, followed him, saying, "Many researchers are suffering from practice-oriented policy and are afraid of concluding and generalizing (their research) with foresight" (ibid., 15).

35 Nikolai Akulov, a professor at MSU, had such a strong hostility toward Semyonov that he sent several letters accusing Semyonov to the Party's Central Committee. The points of his accusation can be known in the most well-formulated letter in 1946 (date – illegible): (1) Semyonov's "monopoly" of the academic posts and the spaces in journals; (2) the mutual supporting relationship with a German scientist, Ernest Bodenstein, for self-advertisement in order to patch up the contribution of a Russian scientists, N.A. Shilov, in the formation of chemical chain reaction theory; (3) concoction of experimental data; (4) failure in linkage of theory with the practice that brought some confusion to the actual sites (RGASPI, F. 17, Op. 125, D. 449, 127–139). The Central Committee organized a commission to settle the dispute by listening to the complaints from both sides. The commission denied the assertion of Akulov and supported Semyonov in conclusion (ibid., 141–152).

36 ARAN, F. 342, Op. 1, D. 45–221, 42.

37 Ibid., 46.

38 ARAN, F. 393, Op. 1, "Predislovie."

39 ARAN, F. 1968, Op. 1, "Predislovie."

40 ARAN, F. 393, Op. 1, D. 50–211, 31.

41 In his letter dated December 1, 1947, to a Central Committee secretary, Aleksei Kuznetsov (1905–1950), Dankov complained about the interference from the side of Frumkin and his colleagues in the publication of his paper written jointly with A.V. Frost to criticize S.Z. Roginskii who belonged to the Frumkin's group (RGASPI, F. 342, Op. 1, D. 45–221, 42).

42 ARAN, F. 393, Op. 1, D. 63–631, 1, 59–149.

43 Ibid., 99.

44 Ibid., 99: As soon as Dankov ended his speech, a researcher, Burshtein (his name, patronymic and other biographical details – unknown) replied to Dankiov, asking, "Doesn't Dankov cite any work by foreign scientists in your works?" (ibid., 119).

45 Ibid., 146.

46 RGASPI, F. 17, Op. 125, D. 618, 163: It is known from Yurii's memoramdum to his father, Andrei Zhdanov, attached to a letter sent to him by a physicist, A.P. Abramovich, to accuse "the monopolist group" (Abramovich attached this letter to his own papers, "Physicists=Monopolists in the light of Philosophical Discussion" and "Lenin and contemporary physics" sent for review to the Party' Central Committee) that Yurii, who was a graduate from the Faculty of Chemistry of MSU, himself was anxious about monopolistic tendency of some physicists such as Kapitsa, Landsberg, Landau, Leontovich, and others (ibid., 1–4).

47 ARAN, F. 1968, Op. 1, "Predislovie": The lack of governability of Frumkin might have been considered; he could not manage his institute well, having left room for the activities of his opponents. However, his fame and authority as an outstanding scientist was not undermined. He maintained a full-time membership of the Academy of Sciences and the post of the chief of the Electrochemical Division of his institute. Later, in accordance with the independence of his division as the Electrochemical Institute, he was appointed its Director (ibid.).

48 ARAN, F. 393, Op. 1, D. 65–211, 221, 14.

49 Vizgin, "Yadernyi shchit . . ." *op. cit.*, in note 16. 1373.

50 Ibid., 1373–1379.

51 E. Pollock, *Stalin and the Soviet Science Wars*. Princeton and Oxford, Princeton University Press, 2006, 87.

52 Ibid., 1375.

53 Ibid., 1375–1379.

54 See, Ya. Rokityanskii i dr., *Sud Palacha: Nikolai Vavilov v Zastenkakh NKVD – Biograficheskii Ocherk. Dokumenty*. Moscow, Academi, 1999: The reason why

Sergei was given such important posts is unknown. It is said that the promotion of Sergei was an expression of Stalin's expiation. Or, it is also said that the political power took advantage of Sergei's "obedience" because of his fear. In any case, his case was a rare one in those days when in many cases almost all relatives of a criminal were also involved in punishment.

55 Yu.I. Solov'yov, "Akademik S.I. Vavilov: Drama russkogo inteligenta." *Voprosy Istorii Estestvoznaniya i Tekhniki*, No. 1, 1999, 136.

56 ARAN, F. 596, Op. 2, D. 25, 1–3. This document lacks its date. However, it can be supposed that it was written in 1943.

57 Solov'yov, *op. cit.*, in note 55. 138: It is said that Komarov was virtually dismissed from his post due to displeasure of the political authority against him because he didn't refer to the meritorious achievement of Stalin in the occasion of the 220th Anniversary of the Academy of Sciences (ibid.).

58 Ibid., 142, 143: On the other hand, Sergei Vavilov tried enthusiastically to protect contemporary science and the interests of scientists against interference from the political power by taking part in the plan to establish the Institute of Cytogeneticss in opposition to the Lysenkoites (later, he expressed his self-criticism), making effort to include the physicists belonging to the Academy's institutes to the "All-Union Conference of the Physics Department Chiefs of Univerisitie and Higher Education Facilities" and rename it "All-Union Conference of Physicists," suggesting Fok, Sergei Frish, and other physicists to take an action to protect contemporary physics, trying, himself, to protect quantum dynamics and the theory of relativity in his speech continued for two hours in the ceremonial session dedicated to "the history of father-land science" of the Academy of Sciences in Leningrad in January 1949 and taking other means (Vizgin, "Yadernyi shchit . . ." *op. cit.*, in note 16. 1378, 1379).

59 Vizgin, "Yadernyi shchit . . ." *op. cit.*, in note 16. 1376.

60 V.D. Esakov, "Mify i Zhizn'." *Nauka i Zhizn'*, No. 11, 1991, 110.

61 Even only in the first three months of 1948, totally seven indictments on the worshipers of foreign science were received by the Party' Central Committee (RGASPI, F. 17, Op. 125, D. 618, 2–161).

62 Vizgin, "Yadernyi shchit . . ." *op. cit.*, in note 16. 1372.

63 RGASPI, F. 17, Op. 125, D. 618, 170.

64 For example, see, F.A. Korolyov, "O metodorogicheskikh oshibakh v knige prof. S.E. Khaikina Mekhanika." *Uspekhi Fizicheskikh Nauk*, Vol. 37, No. 3, March 1949, 388–394; This paper shows so straightforward that the "patriotic and materialistic physicists" were not qualified to conduct philosophical debates. However, a Fizfak professor, Yakov Terletskii (1912–1993), is an interesting exception; he devoted himself so enthusiastically to the philosophical interpretation of contemporary physics in search of "physics" consistent with Marxist philosophy. See, a series of his philosophical articles published in the journal, *Voprosy Filosofii*; "Problemy razvitiya kvantvoi teorii" (No. 5, 1951, 51–61), "O soderzhanii sovremennoi fizicheskoi teorii prostanstva i bremeni"(No. 3, 1952, 191–197), "Ob izlozhenii osnov spetsial'noi teorii otnositel'nosti" (No. 4, 1953, 207–212), "O vzaimoprevrashchaemosti elemental'nykh chastits" (No. 2, 1956), etc.

65 About one year after the discussion in that institute, the situation was radicalized. Yurii Pobedonostsev, a member of the Artillerist Academy, M. Tikhomirov, its corresponding member, and N. Chernyshev, Lieutenant Colonel in technical service, set their co-authored article, titled, "Allegoriya Mistera Vailda i radost' Professora Sokolika (The Allegory of Mr. Wild and the Joy of Professor Sokolik)" to the editorial board of *Izvestiya* for publication. They complained that in the writing by "Mr. Wild (the details – unknown)" on the history of rocket technology, there

was no reference to the priority of Russian scientists such as Konstantin Tsiolkovskii (1857–1935) and all achievements in this field were referred to the Western scientists. And then they accused Sokolik for his uncritical attitude toward Wild's position in introducing his writing to the Soviet Union. In a meeting of the Academic Council of the Institute of Chemical Physics, called for the discussion on this matter, Sokolik was asked for his "self-criticism." However, he refused doing so, saying, "The writings having the highest theoretical value in foreign literature are not published generally due to the well-known reason. Only poor materials like sawdust exhausted in the scientific and technological creations are permitted for the publication. The editors responsible for the edition of the collections must engage not in such sawdust, but in the original materials which have been known valuable in scientific and technological aspects. Hence, I thought that I, as an editor, should bring some profit to the Soviet engineers in the way to give an opportunity for them to be acknowledged with the concrete technological materials, even if they have only second-rate value. Personally I don't consider this writing as a history of rocket engines. If it had given such impression, I would not have inserted it to the collection. Because, it would not attract any interest. I suppose that we have some writers who can and ought to write on the history of rocket engines" (ibid., 113, 114). In this way, this man of firm character denied necessity of his "self-criticism" even in the last opportunity (ibid., 146). Subsequently after Sokolik, a researcher, Voinov (his first name and patronymic – unknown), said, "The article on *Izvestiya* showed us how to introduce foreign literature. We must take it a serious signal for all of us" (ibid., 114).

66 Even when Sergei Suvorov, Deputy Chief of the Central Committee's Propaganda and Agitation Section, made a proposal about the inevitability of criticism against Neo-Kantian view of the bourgeois nuclear physicists in his letter dated July 14, 1947 to Zhdanov, a Politburo member, Suvorov requested Zhdanov for the careful treatment "not to bring harmful effects to physics itself" (RGASPI, F. 17, Op. 125, D. 492, 16, 17).

67 In addition to this, we must take into consideration the ambition of the young, newly promoted party apparatchiks like Yurii Zhdanov to make striking achievements.

68 Yu.I. Krivonosov, "Okolo atomnogo proekta." V.P. Vizgin (Ed.), *Istoriya Sovetskogo Atomnogo Proekta: Dokumenty, Vospominaniya, Issledovaniya*, Vypusk 2. St. Petersburg, Izd-vo Russkogo Khristianskogo Guamnitarnogo Instituta, 2002, 354–362.

69 Krivonosov, "Srazhenie na . . ." *op. cit.*, in note 10. 29.

70 Vizgin, "Yadernyi shchit . . ." *op. cit.*, in note 16. 1369.

71 V.A. Fok, "Problemy atomnogi fiziki (rukopis')." Sankt-Peterburgskii filial Arkhiva Rossiiskoi Akademii nauk (PF ARAN), F. 1034, Op. 1, D. 356, a clipping.

72 V.A. Fok, "Osnovye zakony fiziki v svete dialekticheskogo materializma (rukopisi)." PF ARAN, F. 1034, Op. 1, D. 370, 34–47.

73 V.A. Fok, "Osnovye zakony fiziki v svete dialekticheskogo materializma." *Vestnik Leningradskogo Universiteta*, No. 4, 1949, 46, 47.

74 A.F. Ioffe, "Bor'ba s idealizmom v fizike: Tezisy doklada na uchonom sovete FTI AN SSSP." PF ARAN, F. 910, Op. 1, D. 202, 1.

75 Yu.V. Goponov, S.K. Kovaleva and A.V. Kessenikh, "Studencheskie vystpleniya 1953 goda na Fizfake MGU kak sotsial'noe echo atomnogo proekta." *Vizgin (Ed.) Istoriya . . . op. cit.*, in note 68. 527–537.

76 Before that, in August 1949, a large-scale recertification of all MSU professors and teachers was conducted. As a result of it, Timiryazev, an active member of "patriotic and materialistic physicists," was compelled to retire from his profession for his advanced age (TsMAM, F. 1909, Op. 2, D. 283, 15–17).

77 Slava Gerovitch, *From Newspeak to Cyberspeak: a History of Soviet Cybernetics*. Cambridge, MA, London, The MIT Press, 2002, 7: For example of such "self-perpetuating" expansion, Gerovitch cites the case of Mikhail Yaroshevskii (1915–2001). Yaroshevskii attacked cybernetics as a "bourgeois" pseudo-science (see, M.G. Yaroshevskii, "Semanticheskii idealism – Filosofiya imperialisticheskoi reaktsii" in Institut Filosofii AN SSSR (Ed.), *Protiv Filosostvuyushchikh Oruzhenostsev Amerikano-Angliiskoi Imperializma: Ocherki Kritiki Sovremennoi Bruzhuaznoi Filosofii i Sotsologii*. Moscow, Izd-vo AN SSSR, 1951, 88–101; (M.G. Yaroshevskii, "Kibernetika: 'nauka' mrakobesov." *Literaturnaya Gazeta*, 5 aprelya 1951, 4). According to Gerovitch, he didn't have a party membership and was once arrested in 1938. After the war, he was demoted to Tajikistan due to his "cosmopolitanism." Therefore, his articles written in such a philosophical offensive must have been a wonder drug capable of reviving his fame. However, according to the memoir of Yurii Zhdanov, Stalin, himself, considered mathematical sciences including cybernetics important for the development of military-related technology (Yu.A. Zhdanov, "Vo mgle protivorechii." *Voprosy Filosofii*, No. 4, 1993, 89). If Yurii's testimony is true, the criticism against cybernetics by some philosophers such as Yaroshevskii was conducted without any direction or support from the top level of the political authority. That is the reason why Gerovitch described such an offensive of the philosophers as "self-perpetuating" expansion. In addition, an epoch-making article titled "Basic Features of Cybernetics," co-authored by Sergei Sobolev, Anatolii Kitov (1920–2005), and Aleksei Lyapunov (1911–1973), was issued in 1955 (S.L. Sobolev, A.I. Kitov and A.A. Lyapunov, "Osnovnye cherty kibernetiki." *Voprosy Filosofii*, No. 4, 1955, 136–155. As for the significance of the publication of this article, see Gerovitch, *op. cit.*, 173–181).

78 Institut filosofii, . . . *op. cit.*, in note 77.

79 ARAN, F. 1922, Op. 1, D. 726, 110–115.

80 ARAN, F. 2, Op. 6, D. 148, 78, 79.

81 I.P. Grinina and S.S. Ilizarov, "Otdel nauki TsK KPSS v period politicheskogo krizisa 50-kh godov." in Institut Istorii Estestvoznaniya i Tekhniki RAN (Ed.), *Godichnaya Nauchnaya Konferentsiya*, 1996g, 79–87.

82 Dismissal of Aleksandrov on March 10 was attributed to his lack of governability. However, we should consider it as an incident related to the establishment of the hegemony of Nikita Khrushchev (K.A. Zalesskii, *Imperiya Stalina: Biograficheskii Entsklopedicheskii Slovar'*. Moscow, VEChE, 2000g, 19): He died suddenly due to a cerebral hemorrhage on June 21, 1961.

83 ARAN, F. 2, Op. 6, D. 194, 24–27.

84 Ibid., 135–146.

85 Ibid., 147.

86 Ibid., 148–151: Another meeting of the Presidium was held in June 17 so as to brush up the draft of their proclamation. In this meeting, a young Presidium member, Mstislav Keldysh, demanded an amendment because of his anxiety that the Presidium might have referred all the errors of the philosophers to Aleksandrov alone (ARAN, F. 2, Op. 6, D. 195, 219–224). On the other hand, in spite of such offensive against them, the Institute of Philosophy didn't cease their "fight" against "bourgeois idealism." They planned to issue a reference book under the title, *Burzhuaznaya Filosofiya i Sotsiologiya v Ere Imperializma* (*Bourgeois Philosophy and Sociology in the Age of Imperialism*). (ARAN, F. 1922, Op. 1, D. 726, 110–113).

87 ARAN, F. 1647, Op. 1, No. 104, 6.

88 P.N. Fedoseev et al. (Ed.), *Filosofskie Problemy Sovremennogo Estestvoznaniya: Trudy Vsesoyuznogo Soveshchaniya po Voprosam Estestvoznaniya*. Moscow, Izd-vo AN SSSR, 1959, 3, 4.

89 Ibid., 573.

Part II

Some aspects of the Soviet Cold War research and developments

3 "Foresight" or "survival"? Rocket development and the Ministry of Armament

What is curious about the activities of Ministry of Armament immediately after World War II is that they had charge of the research and development of rockets with liquid propellant fuel in spite of their lack of experience, and that they started their project with the mere copying of the German rockets in spite of objections from the Soviet leading engineers. Here, the author tries to make clear the business factor for the Ministry to make such choices.

1 Wartime expansion of the people's Commissariat of Armaments and the conversion after the war

1.1 Wartime expansion of military production

In 1940, the total sum of the orders by the People's Commissariat of Defense was 30.9 billion rubles. Of them, 32.1% were allotted to the People's Commissariat of Ammunition, 23.5% to the People's Commissariat of Aviation Industry, 7.6% to the People's Commissariat of Medium-size Machine-building Industry, 2.9% to the People's Commissariat of Chemical Industry, 2.3% to the People's Commissariat of Normal Machine-Building Industry, 1.5% to the People's Commissariat of Ship-building Industry, 2.9% to the People's Commissariat of Defense itself, 16.2% to other People's Commissariats, and the remaining 16.2% were allotted to the People's Commissariat of Armaments.[1]

The first step taken by the Soviet government for the reinforcement of defense industry was the proclamation of the Council of People's Commissars dated August 29, 1939, "On the balance in distributions and planning of high-quality steel and ferroalloy." Soon after this measurement, the resolution of the Party's Politburo, "On the re-equipment of the existing aviation factories and the construction of new factories," announced the initiation of the mobilization of industrial potentials for the war by declaring the construction of nine new factories and the conversion of nine other factories to military production. At the same time, approximately 60 civil industrial factories were transferred to the People's Commissariat of Aviation Industry. In addition, the orders for munitions placed to the enterprises of civilian industry were rapidly increased. In June 1940, the Politburo decided to mobilize two tractor factories in Chelyabinsk and Stalingrad (now, Volgograd) for the production of T34 type tanks, Around that time, the Ural Heavy Machine-building Factory, the Ural Freight Wagon Factory and the machine-building factories in Novo-Chelkassk, Novo-Kramatr, and Votkinsk went for the production of gun barrels and clad metals (a metal sheet on the surface of which another sort of metal was covered: bimetal). The number of representatives of the military forces dispatched to various factories for quality control of munitions was increased (by 1.5 times) to 20,281 personnel in a couple of years from 1938.[2]

After the break-out of the war between the Soviet Union and Nazi Germany, such conversion of industrial potentials to military industry was accelerated. On June 23, 1941, the Politburo approved the governmental decision to transfer many civil industrial enterprises to the People's Commissariat of Ammunition. Eventually the People's Commissariat came to manage approximately 600 enterprises in addition to 65 enterprises that belonged to that Commissariat from the outset.[3] According to the instruction of the Presidium of the Supreme Council dated September 11, the People's Commissariat of Tank Industry was newly established on the basis of various departments of People's Commissariats. This Commissariat affiliated 27 enterprises, including the largest industrial facilities in the Soviet Union like the Kirov Factory for tractor manufacturing in Chelyabinsk, the Factory No. 112 "Krasnoe Solmovo" for ship-building, the Ural Heavy Machine-building Factory, Stalingrad Tractor Factory, and others. These enterprises commanded 218,300 personnel and provided diesel engines, cannonballs, bodies of submarines, guns equipped to combat planes and bombers, and others besides tanks.[4] Subsequently on November 21, a measure was adopted to reorganize and shift the People's Commissariat of Normal Machine-Building Industry to the new People's Commissariat of Trench Mortars and Rocket Guns.[5]

Even before the outbreak of the war, the People's Commissariat of Armaments had affiliated 43 factories (286,000 personnel and 70,800 machine tools): 14 gun factories, nine arms factories, eight factories providing cartridges for guns and arms, and 12 optical weapon factories.[6]

"Evacuation to the East" was initiated by the joint resolution dated August 16, 1941, of the Council of People's Commissars and the Party's Central Committee to move the factories located in European Russia relatively near to the safe zone of the territory so as to secure industrial potentials. Immediately, it was decided to move a total of 1,360 enterprises: 455 enterprises to Ural, 210 to Western Siberia, 250 to Kazakhstan and Central Asia, and so on.[7] "Evacuation to the East" was promoted drastically under the supervision of "Operational Bureau" affiliated to the provisional supreme war-commanding organization, i.e., the State Defense Committee. The "Bureau" then consisted of political leaders such as Vyatislav Molotov (1890–1986, Chair), Georgii Malen'kov (1902–1988), and Anastas Mikoyan (1895–1978). Later, on December 8, 1944, the "Bureau" also included Lavrentii Beriya (1899–1953, Chair), Malen'kov, Mikoyan, Nikolai Voznesenskii (1903–1950), and Kliment Voroshilov (1881–1969).[8] In October of this year, 324 of 501 enterprises affiliated to the People's Commissariats of Aviation Industry, Tank Industry, Armaments, Trench Mortars and Rocket

Guns, and Ship-building were under evacuation, i.e., only 177 enterprises continued their normal industrial activities.[9]

As for the People's Commissariats of Armaments, 32 of the 58 enterprises were on the way to the East. "Evacuation to the East," which composed of large-scale transportations and installations of industrial equipment to local manufacturing facilities for operation on the spot, with the full use of trains and ships, was often accompanied with serious difficulties according to the variety of local conditions at the location. For instance, the equipment of the Factory No. 60 of the People's Commissariat had to be dispersed to various destinations, i.e., to Factory No. 540 in Irkutsk, Factory No. 543 in Kazan,' Factory No. 541, Factory No. 545 in Chkalov, and Factory No. 537 in Kirov (all these factories seem to have been renamed on the occasion of evacuation), and the Repair Works of the People's Commissariat of Local Industry of Kyrgyz Republic.[10]

1.2 Postwar conversion to civilian production and "the Crisis" of the Ministry of Armament

The annual average expenditure for national defense reached 58.2 billion rubles, which was 50.8% of the national budget during 1941–1945. The ratio in the national income reached 33% in those days. However, the defense expenditure rapidly declined to 12.8 billion rubles in 1945, which is the year the war ended, and then further declined to 7.3 billion rubles in 1946 and 6.6 billion rubles in 1947 and 1948.[11]

When the enterprises of non-military industry were reorganized for military production, they were obligated to continuously provide goods of civilian use. However, during the war, the production of military products pressured the production of civilian products. The ratio of the production of civilian products was reduced to 7.1% in the enterprises affiliated to the People's Commissariat of Aviation industry, 6.4% in the People' Commissariat of Ammunitions, and 7.1% in the People's Commissariat of Tank Industry.[12] The shortage of civilian goods was compensated, to a large extent, by the goods exported from the Allied Powers and the goods provided by them under the "Lend-Lease" program, a project to supply industrial goods to the Soviet Union as aids. In particular, "Lend-Lease" was a huge operation to transport 16 million tons of goods, for which a total of 2,660 transport ships were used. The number of munitions and weapons supplied to the Soviet Union under this program was huge; the ratio of the usual infantry outfits supplied reached 7.4%, that of tanks and self-propelled guns was 13.4%, and that of warplanes was 20.5%.[13] The total sum of "Lend-Lease" reached 106 billion U.S. dollars, which covered 18.9% of the national income of the Soviet Union. Moreover, the tremendous amount of civilian goods and industrial materials provided under this program should be taken into consideration. According to the data of the First Department of the State Planning Committee (Gosplan), at least 55% of the trucks and cars, 20.6% of tractors, 23.1% of metal-cutting machine tools, 42.1% of locomotives, 40.8% of aluminum, 19.3% of zinc, 25.3% of nickel, 37% of mercury, 99.3% of tin, 56.9% of cobalt, 67.1% of molybdenum, 24.3% of anti-rust steel, 18.1% of benzene for airplanes, 22.3% of ethyl alcohol, and 38.2% of glycerin were supplied under the "Lend-Lease" program during the period of 1941–1945.[14] The end of the war also meant the termination of the "Lend-Lease." In order to satisfy the surplus demands for civilian goods, a large-scale operation was carried out to transport various kinds of industrial equipment to the territory of the Soviet Union from over 5,500 "hostile" enterprises in the territories occupied by the Soviet Army (in eastern Germany, Poland, some parts of Austria, Hungary, Northeast China, and Korea).[15] This operation was conducted by the Commission for the Transfer of Equipment of Japanese and German Firms, established within the Council of People's Commissars with Avraamii Zavenyagin (1891–1956), then a deputy of People's Commissar of Interior Affairs, as Chair.[16] Later, in December 1947, a new governmental agency, the State Commission of Supply (Gossnab), was established on the basis of some departments separated from Gosplan to take charge of the distribution of the acquired equipment. At the same time, the State Commission for Introduction of New Technology to National Economy (Gostekhnika) was established to manage a great deal of foreign patents provided by the United States during the war and requisitioned from Germany and to encourage their introduction to various factories. According to the proclamation No. 600–240 of the Council of Ministers dated February 13, 1949, regarding 225,000 patents managed by Gostekhnika, of them 45,000 were German patents registered in 1939–1940, 50,000 also were German patents registered thereafter (including those that were not registered yet officially), and 130,000 came from the United States.[17]

Beside the use of such "Trophies," it was intended to reduce the military industry (that was excessively inflated during the war) to the rational scale. By Decree of the State Defense Committee of May 20, 1945, the People's Commissariats of Ammunitions, of Trench Mortars and Rocket Guns, and of Tank Industry were liquidated. According to this decision, the factories manufacturing tanks and diesel engines were transferred to the People's Commissariat of Means of Transportation, the factories producing cartridges and grenades, the People's Commissariat of Agricultural Machine-building, and so on. Subsequently, a proclamation titled "On the reorganizing measures of industry accompanied with the reduction of weapon production," issued by the State Defense Committee on May 26 of the same year, instructed three People's Commissariats, including the People's Commissariat of Armaments, to boldly pursue the conversion to peaceful production; the People's Commissariat of Aviation Industry was obligated to set aside some portion of their production capacity to the production of energy facilities, electrical equipment, autobus, motorcycles, and others. Similarly, the People's Commissariat of Armaments went for the manufacturing of optical apparatuses, winches, transferring devices (telphers and others), mining machines for coal and petroleum, and others. The ratio of military products in the total output of five ministries (People's Commissariats were renamed Ministries March 15, 1946) including the Ministry of Agricultural Machine-building, the Ministry of the Means of Transportation, the Ministry of Communication Device Industry along with the two above-mentioned ministries diminished to 34.5% in 1945.

In the second half of 1945, this ratio drastically reduced to 10.8% in 1947. In contrast, the proportion of civil products in these five ministries increased to 49.1% in the second quarter of 1947 (the difference between the proportion of military products and that of civil products was that of dial use products or intermediate goods).[18] At a meeting on April 22, 1946, of the Collegium of the Ministry of Armament, Minister Dmitrii Ustinov (1908–1984) made a speech: "The problem of capital construction is a matter of utmost importance concerning the conversion of industry to the new line – the line for peaceful life."[19] The ratio of production of commercial goods in the total output of the Ministry increased from 13.7% in 1943 to 47.0% in 1946 (see Table 3.1).[20] The number of factories under jurisdiction of the Ministry decreased from 64 in its peak to 47 in those days. These factories provided 223 different industrial products, including various new products.[21] In spite of the severe fall down of the output in accordance with the end of the war, the activities of research and development in the research institutions of the Ministry and the design bureaus affiliated to various factories did not decline owing to the appearance of such new products. "The total output" of such research and development activities, which was represented in monetary units (at par in 1940) according to the rules implemented by the Ministry of Armament, remained 88.9% in 1945 (84.2% in 1944) in comparison with the previous years respectively.[22]

The State Optics Institute,[23] which was established by the Tsarist government to deal effectively with the shortage of optical glass during World War I, played a major role in introducing modern physics into the Soviet Union. On a trial basis, it manufactured 35 items, 206 samples of optical apparatuses, 14 sorts, 133 samples of camera lenses; prepared 72 reports (on 306 instances) and 243 designs with estimates; and delivered them to various related factories and research organizations. It also issued 171 scientific papers and dispatched their colleagues to 54 factories for technical assistance.[24] Although there is no evidence

Table 3.1 Output of the products prepared by the Ministry of Armament in 1946 (million rubles. Converted into par in 1926–1927)

	Previous fiscal year	*Plan*		*Results*	*Ratio to the previous year*	*Ratio to the plan*	
		Total sum of 4 quarters	*A year*			*4 quarters*	*A year*
Commercial products	9350.0	5161.8	5328.5	4770.6	51.0%	92.4%	89.5%
Military use	8066.6	2892.1	2855.0	2527.1	31.3%	87.4%	88.7%
Civilian use	1283.4	2269.7	2473.0	2243.5	174.8%	98.8%	90.7%
Total	9395.8	5422.9	5596.5	4962.4	52.8%	91.5%	88.7%

*Numbers are not consistent in many cases. There might be some errors in calculation and/or some products unseen in this table.

Source: RGAE, F.8157, Op. 1, No. 2542. 36.

on the use of "Trophy" goods in the research and development activities, it is supposed that German technology was widely used in such active research and development. A report tells, "The operations (concerning development and design – *Ichikawa*) were carried out in Moscow, in various operational units and in Germany (for the purpose of acceleration of the design works and practical resolution of the various problems that often occur in the processes of the design and constructive works)."[25]

The measure to convert industrial production to peaceful one was sustained for a long period. As seen in Table 3.2, even in 1950, the Ministry continuously provided a large variety of industrial goods for civilian uses, including various optical apparatuses.

The conversion, however, led the Ministry to a serious financial crisis. On December 10, 1945, the Industrial Bank sent a letter to Minister Ustinov cautioning: "An analysis conducted by the USSR Industrial Bank on the reports of the contracting organizations of the People's Commissariat of Armament showed extremely unsatisfactory results of financial and economic activities of the trusts for three quarters of 1945."[26] In addition, in a resolution of the Collegium of June 17, 1946, the Ministry itself admitted:

> Having heard the reports on the financial situation of the Ministry of Armaments, the Collegium of the Ministry of Armaments states: The financial situation of the industry managed by the Ministry of Armaments in May 1 is extremely serious. The debts owed by the Ministry of Armaments to the State Bank and the suppliers exceed the amount of funds the Ministry of Armaments now has and the unsettled payments the customers have to pay to us by 525.3 million rubles.[27]

Table 3.3 shows the huge amount of debts owed by the Ministry of Armament. Generally speaking, the development and manufacturing of a new product inevitably needs a sort of "pregnant period" from the input of additional expenditure and additional supply of labor power to the output of economic achievements. It involved a series of engineering operations including the introduction of new equipment, its installment and adjustment, and completion of the sample to a final product. Therefore, it is taken for granted that such innovation accompanies a temporary loss. As seen in the proviso to Table 3.3, some amount of loss had been foreseen. However, it should be noted that the factors accounting for the losses exceeded the anticipated estimate. Concerning the factor seen in Table 3.3, i.e., "fulfilled or unfulfilled supply of products and materials," the memorandum attached to the annual accounting report for the year 1946 mentions some ordinary or heteronomous factors common to the Soviet industry as a whole such as shortage of coal and other fuels, instability of the supply of electricity, a sharp decrease of the orders for the armaments and munitions, compulsory supply of labor power for agriculture and civil engineering works,[28] and disordered coordination in the distribution of the supplies of the materials and parts among the enterprises. Many of these factors were related to the large confusion regarding the Soviet national economy immediately after the war. The memorandum also

Table 3.2 The list of products for civilian use prepared by Ministry of Armament in 1950

• 27 items of open-hearth-furnace steel and electric-furnace steel, • Metal cutters, • Screw cutters, • Precision lathes, • Turret lathes, • Milling cutters, • Gear cutters, • Planers, • Polishers, • Specialized metal cutters, • Presses and other plastic working machines, • Water turbines, • 75 horsepower steam engines, • Oil engines and the parts, • 2 horsepower stationary engines, • 50 horsepower vessel engines, • 5 types of electric motors, • Electric cranes, • Hammer-drills, • Core drills, • Ore loading machines, • Parts for hammer-drills, • Boring mills, • Excavators, • Pumps for deep excavation, • Pumps for dirt, • Adjusters for boring mills, • Adjusters for petrol equipment, • Drills and pipes for pumps. • SKN-3 type bailing pumps, • SKN-6 type bailing pumps, • Turbine boring mills, • Extenders (*content – unknown*), • Gal chains (*content – unknown*), • Keystone chain, • Chains for presses, • Electric saws and the parts, • Chains for saws, • Chains for pulleys and bearings, • Spinning machines and other textile equipment and the parts, • Equipment and parts for knitwear industry, • 5 sorts of shoemaking machines, other equipment and parts for shoemaking works, • Rolling mills, other equipment and parts for metallurgical factories, • Steel-made armatures, • 3 types of excavators, • Steam locomotives for narrow gauges, • Drain pumps with slide valves, • Parts for repair of transportation machines, • 4 types of electric transporters, • 2 types of trucks for cranes, • Compressors, • Parts for electric power stations, • Fixtures clamps, • Files, • Chucks for hand-drills, • Chucks for lathes, • Universal dividing head, • Seismological observation devices, • Calcium carbide, • Film projectors, other optical devices and parts, • Microscopes, • Microscopes with universal measuring instruments, • Metallurgical microscopes, • Mineralogical microscopes, • Spectrographic microscopes, • Multi-purpose microscopes, • Biological microscopes, • Electron microscopes, • Spectrographs, • Universal tachometers, • Projectors, • 3 types of spirit levels, • 6 types of theodolites, • Optical angle measuring instruments, • Sensitometers, • Monochromatic light spectroscopes, • Vertical measuring instruments, "Abbe," • Large kinreter (*content – unknown*) with measuring boards, • Micro-photometers, • SpectrographISP-51, • Microwave interferometers, • Horizontal measuring instruments, "Abbe," • Polariscopes, • Drovishev-type slide rules, • Dzhenevskii-type slide rules, • Optical dividing heads, • Magnifying glasses for both eyes, • Tebrer-type pipes (*content – unknown*), • Covers for lenses without aberration, • Measuring instruments (*content – unknown*), • Double spectrographic projectors, • Sphero-meters, • Spectrographic photometers, • Dubosoc color phase meters, • Erico color phase meters, • Projector vision scope meters, • Sali-type hemoglobin meters, • Millar shells for photometers, • Circular polariscopes, • Lumps for photometers, • Loupes for single eye, • Loupes for both eyes, • Capillary scopes, • Rollnet tri-view mirrors, • Light adaptation meters, • Gear blocks DZB, • Gas densimeters, • View finders, • Light transposers, • 2 types of magnifying glasses, • Stephan measuring compasses, • Edison electric sockets, • Electric switches, • Hunting loupes, • Small caliber guns, TOZ-8 and TOZ-11, • 2 types of cartridges for hunting rifles, • Long-type ammunition cartridges, • Motorcycles, K-125 and Izh-330, • Springs for gramophones, • 3 types of axes and hatchets, • Luxury glass tableware, • Metal-fabricated beds, • Knives for restaurants and cafes, • 2 types of electric irons, • 20 sorts of parts for agricultural tractors (axles, cylinders and others), • 3 sorts of parts for Universal-type tractors (clutches, hubs, control stick gears and others), • 14 sorts of parts for STZ-NATI (*content – unknown*), • 3 sorts of parts for S-80 (*content – unknown*), • Materials for agricultural machines (*content – unknown*), • Evalga chains, • Jackson chains, • Headless rivets for drums.
*Besides, – totally 14 illegible items.

Source: RGAE, F.8157, Op. 1, No. 2578. 31–47.

Table 3.3 Profits and losses of the Ministry of Armament in 1946 (million rubles)

Factors for profits or losses	*Profit*	*Loss*
a) Fulfilled or unfulfilled supply of products and materials	–	363.3
b) Withdrawal of orders	–	142.8
c) Suspension of operations in the factories for more than 15 days due to pause of the supply of electricity and/or fuel	–	43.9
d) Utilization for common facilities for public life	–	39.4
e) Inevitable disasters	–	9.8
f) Relief of debts which passed the prescriptions for prosecution or matured not being expected to be paid	23.5	–
g) Factors caused by previous operations	–	16.3
h) Others	–	11.5
Total	23.5	631.7

*Total losses reaching to 341 million 600 thousand rubles had been previously estimated.

Source: RGAE, F.8157, Op. 1, No. 2542. 77.

mentions other factors related to the conversion to the peaceful production, like increased orders for products that could not be settled smoothly into mass production regime in contrast with the previously mass-produced armaments and munitions, difficulties faced by various factories in response to various unplanned additional orders for small amounts of goods,[29] and high losses in the production processes and inferior goods as products. Referring to the factor mentioned last, the factories affiliated to the Ministry of Armament could not compete efficiently in the aspects of experience and industrial equipment well-adjusted for mass production with the factories affiliated to other ministries, which had engaged in the production of civilian goods from the outset.[30] The Minister of Finance, Arsenii Zverev (1900–1969), inquired the Bureau of the Council of Ministers to examine the financing activities of the Ministry of Armament to cover the deficit in the special revolving fund in July 27, 1947.[31] This issue was deliberated at length in the Bureau meeting of December 4. For this meeting, the Ministry of Finance prepared a material, which consisted of the following points. (1) The Ministry of Armament then owed 2,243 million rubles of expired debts to the State Bank, 321 million rubles to the suppliers, 4 million rubles of unpaid wages to the employees. (2) The Ministry has 91 million rubles of reserved materials as of October 1 that exceeds by far the normal scale of reserves (239 million rubles of materials were introduced to production). (3) The total amount of debts is 435.2 million rubles. (4) The loss made by unproductive expenditure for nine months of 1947 reached 40.5 million rubles and the loss by inferior products reached, surprisingly, 115.6 million rubles.[32]

After the abolitions of three People's Commissariats of Ammunitions, Tank Industry and Trench Mortars and Rocket Guns, the Ministry of Armament remained the only ministry with the profile of managing military industry. In reality, however, the Ministry was losing its core nature. The conversion to civilian industry that was sought for to find a way out of difficulty ended in failure. In addition, it meant the loss of its profile or its identity as a ministry managing military industry. Unlike the People's Commissariat of Tank Industry and Trench Mortars and Rocket Guns that had served as provisional organizations, the People's Commissariat of Ammunitions had a tradition since its establishment as the People's Commissariat of Defense Industry that was separated from the People's Commissariat of Heavy Industry in 1936 (the People's Commissariat of Defense Industry was divided to the People's Commissariats of Ammunitions and of Armaments in 1939). There was a severe situation for the Ministry of Armament in which the Ministry was anxious about its own abolition since even the People's Commissariat of Ammunitions which had a tradition was abolished.

2 German V-2 rocket and its reproduction by the Ministry of Armament

2.1 *The Soviet rocket technology and the investigation mission dispatched to Germany*

The Research Group on Rocket Engine (*Gruppa izucheniya reaktivnogo dvigatelya*; GIRD) affiliated to the Society for the Assistance to Defense and the Construction of Aviation and Chemical Industries (*Obshchetvo sodeistviya oborone i aviatsionno-khimicheskomu stroitel'snvu*; OSOAVIAKhIM) headed by Fridrikh Tsander (1887–1933) gathered such talents like Sergei Korolyov (1907–1966) and others and engaged in the development of rocket engines. They developed the rockets equipped with liquidated fuel and liquidated oxidizer, GIRD-09 (liquidated oxygen and benzine), and, successively, GIRD-10 (liquidated oxygen and alcohol) as early as in the second half of 1933. That group was combined with the Gasdynamics Laboratory (*Gazodinamicheskaya laboratoriya*; GDL) into a single organization, the Reactive Scientific Research Institute (*Reaktivnyi nauchno-issledovatel'skii institut*; RNII) under the People's Commissariat of Heavy Industry. Under the leadership of Ivan Kleimenov (1899–1938), Georgii Langemak (1898–1938), Korolyov, and others enthusiastically worked for the development of rocket ammunitions, M-8, M-13, and others. This resulted in brilliant military gains in the air raids at the battle of Khalkhin-gole (Nom-un Khan, in Japan) in 1937. Furthermore, they succeeded in the development of ground-to-ground rocket guns, types 212, 216, 217, air-to-air rocket guns, and type 201. However, their activities suddenly piled up in 1937–1938. Kleimenov and Langemak were shot to death, and Korolyov, Valentin Glushko (1908–1989), and others were sent to isolated camps, implicated in the downfall of a marshal, Mikhail Tukhachevskii (1893–1937), who was regarded as a patron of RNII.[33]

Thereafter, the Third Scientific Research Institute of the People's Commissariat of Ammunitions continued their own research and development on rocket guns. With aggravation of the war with Nazi Germany, the tactical significance of rocket ammunitions and armaments was enhanced. The State Institute for

Reactive Technology, newly established in July 1942 as a research agency affiliated directly to the Council of People's Commissars (transferred to the People's Commissariat of Aviation Industry in February 1944), played the main role in the rocketry development efforts. In addition, the First State Central Scientific Research Institute of the People's Commissariat of Agricultural Machine-building Industry was in charge of the development of rocket guns with gunpowder. The First Scientific Research Institute of the People's Commissariat of Aviation Industry was into the development of rocket engines with liquidated fuel, and the Design Bureau of the Factory No. 88 affiliated to the People's Commissariat of Armaments was into anti-aircraft rocket guns.[34]

Korolyov and Glushko were released due to the aggravation of the war and compelled to engage in the research of reactive engines in the Joint Design Bureau of the Factory No. 16 of the People's Commissariat of Aviation Industry. Glushko led the Second Design Bureau, whereas Korolyov devoted himself in the research of jet engines for aircrafts as the leader of the Fifth Group.[35]

Soviet engineers had accumulated knowledge through their own practice in the field of rocketry technology until the end of the war. They were also interested in German rocketry, especially in liquidated fuel for V-1, V-2 missiles. The Technical Commission on Reactive Armaments was established within the system of the Military Administration in Berlin in May 1945.[36] General Lev Gaidukov (1911–1999) was appointed as Chair; Anatolii Semyonov and Aleksandr Mrykin (1905–1972) as the Deputies; and Yurii Pobedonostsev (1907–1973), Vasilii Budnik (1913–2007), and other engineers as members. They arrived at Berlin on May 24, 1945, and set to work to replicate ten A-4 type rockets utilizing the materials and the parts left behind in "cooperation" with German engineers. In August, Korolyov, Glushko, Nikolai Pilyugin (1908–1982), Vladimir Barmin (1909–1993), Vasilii Mishin (1907–2001), N. Ryazanskii, Vasilii Kuznetsov, Boris Chertok (1912–2011), and Nikolai Kuznetsov (1911–1995) were newly dispatched and joined in that commission. In this commission, Budnik and others were mainly in charge of restoration of the lost technological documents.[37]

Meanwhile, the People's Commissariat of Aviation Industry also dispatched Prof. Genrikh Abramovich (1911–1995), a deputy of Director of the First Scientific Research Institute of the Commissariat in the mission led by General Nikolai Petrov (1894–1977), to Germany and made him prepare a report of his investigation on the actual situation of rocket building in Germany. According to his report, as the Red Army approached, large-scale exodus of personnel and machinery was organized from the center of German rocketry development, Peenemunde, which had about 150 buildings, 11 launch pads in the area of 80 km^2. However, only two oxygen works, a power station, a few fuel containers and oxidizer containers, and unconnected materials and parts were left behind there when the Soviet troops advanced into the region. Moreover, since most of the qualified scientists and engineers were evacuated to Thüringen district, the core members such as von Braun surrendered to the U.S. troop; thus, only the "second-class" staffs and workers were left.[38]

However, the German scientists and engineers captured by the Soviet troop included Kurt Magnus (1912–2003), a recognized authority in gyroscope study who later wrote a detailed memoir; Franz Lange, a specialist in radar research; Welner Albring (1914–2007), a specialist in aerodynamics; and Hans Hoch, a specialist in automatic control system. Except for Helmut Gröttrup (1916–1981), who would take the role of a leader of German staffs later, none of the others had worked with von Braun and came into contact with rocket technology for the first time in their lives in "Nordhausen."[39]

In August 1945, an interdepartmental commission, which consisted of representatives of some People's Commissariat in charge of military industries and the Main Directorate of Artillery of People's Commissariats of Defense, was held chaired by Gaidukov, who had returned from Germany. The commission confirmed the following urgent tasks to be conducted in Germany: (1) collection and translation of the materials on German rocket technology; (2) establishment of a new specialized research institute in Nordhauzen district; (3) establishment of five technological bureaus and design bureaus in the factories in Nordhausen; and (4) reproduction and test of seven rockets, V-2, based upon the parts left behind (four of them had been reproduced and tested, the other three were sent to Moscow at that moment). Then, the number of "German collaborators" was approximately 1,200.[40]

In July–August 1945, Korolyov and Glushko were dispatched to Mittelberg Factory in Nordhausen in the eastern part of Germany for the reproduction of the German rocket, A-4. In addition to the establishment of their "third factory" in Klein Bodungen, they established "Rabe (abbreviation of *Raketenbau*)" headed by Chertok in Bleicherode and requisitioned an underground factory for liquid oxygen and an engine test site located in Rösten in South Thüringen, attempting to reuse them under the leadership of Glushko. Those facilities were integrated into "Nordhausen" Institute by the end of May 1946. Gaidukov was newly appointed as the Director and Korolyov as the Chief Designer. The first factory in charge of reproduction of the main bodies of rockets, the second factory "Montanya" in charge of engines, and the fourth factory in charge of the development of control systems were established and attached to the Institute. Thus, the Soviet engineers began to reproduce German V-2 rocket.[41]

What sort of technological information did the Ministry of Armament obtain through such activities in Germany? A report dated April 29, 1946, of Pyotr Kirpichnikov (1903–1980), Vice Chairman of State Planning Committee, to Beriya states that the Ministry of Armament obtained (1) an actual object of *Taifun*, unguided anti-aircraft missile (liquid fuel) and the related technological documents; (2) an actual object of *Wasserfall*, radio-guided anti-aircraft missile (with incomplete electrical devices) and the related technological documents (with incomplete detailed drawings); (3) an actual object of *Reintochter*, radio-guided anti-aircraft rocket gun (gunpowder propulsion) without radio apparatus; and (4) the technological documents entirely lacking detailed drawings, an incomplete actual object, V-2, radio-guided rocket and the technological documents entirely lacking detailed drawings and others.[42] Not only the Ministry of Armament but also the former People's Commissariat of Ammunition collected some technological information. Another report of Kirpichnikov to Beriya enumerates the technological acquisitions possessed by the Ministry of Agricultural Machine-building which took over the duties of former People's Commissariat of Ammunition; (1) two actual objects of V-2, the parts for 9–70 missiles of V-2 type, HS-294 artillerist rockets, HS-293A guided rockets, Fritz-X guided rockets, six actual objects of Henschel-293 cruise bomb (lacking of radio apparatus), and the actual objects of solid fuel for 150 rockets; (2) the technological documents such as the photo-album of V-2 and others; and (3) 1,592 items of equipment carried out of an underground factory of *Mittelwerke* and 450 freight wagons of equipment being carried from the company, *Rheinmetal-Borsig*. It was also reported that 29 persons were engaged in the research on V-2 within that Ministry.[43]

What is interesting here is that Gosplan recommended Beriya to charge the Ministry of Armament to reproduce liquid-fuel missiles, V-2, in spite of relatively more actual objects and parts being possessed by the Ministry of Agricultural Machine-building. The latter was recommended to charge research and development of solid fuel missiles, when Gosplan allocated the works concerned to the reproduction of German missiles among Ministries based upon the accumulation of technological information obtained by each of them.[44]

In the winter of 1945, the Ministry (then, People's Commissariat) of Armament set to work on the organization of the Scientific Research Institute No. 88 (NII-88), which would play a central role in the reproduction of V-2. According to the order of People's Commissar of Armament No. 463 ss. issued on November 30, 1945, the Special Design Bureau, which consisted of 250–300 engineers as full members and was led by Chief Designer P.I. Kostin, was established and attached to NII-88. In the early days, eight core designers headed by Senior Designer P.N. Baikovskii recruited young specialists and engineers just discharged from military service. The work space of the new Institute such as offices for designers, experimental works, laboratories, and others were more equipped, making use of the spare space of Factory No. 88. In the spring of 1946, NII-88 was finally established with Aleksandr Kalistratov as the Acting Director and Kostin as the Acting Chief of Design Bureau. The Institute had 150 staff members as of May 25: eight managing staff members, 15 senior designers and design engineers, 59 engineers, 44 technicians, and others.[45]

2.2 *Evaluation of the German missile, V-2*

To study the rockets of the Third Reich "V-2" with respect to scientific and technical aspects is one thing, but to rely on it as the newest effective weapon is a completely different matter. A high evaluation came to be given to long-range missiles only after the second half of the '50s when the head of the missile was equipped with a nuclear charge. The economic performance of a missile with a warhead unit equipped only with an ordinary explosive is very low. Despite the enormous influence of this weapon on the subsequent strategy and tactics of military operations, the rocket at that time was still considered to be unpromising.

It was in mid-1950s that the rocket development in the Soviet Union was combined with nuclear weapon strategy. A project to develop the missile, R-5, which would have a range of 1,200 km and weight of 28 tons (with a warhead of 1.5 ton), was planned since 1951, when the design of R-3A was finished and completed after three stages of test research conducted from March 1953 to February 1955. In the final stage of test research on R-5, a proposal to equip it with a nuclear warhead was presented and adapted. The physicists who were playing the leading role in the nuclear weapon project like Yurii Kahriton (1904–1996), Arkadii Brish (1917–2016), Andrei Sakharov (1921–1989. A Nobel Peace Prize Laureate in 1975), and others engaged in the design of such a warhead. The newly developed strategic missile, R-5A, was tested in January–February 1955 and positioned to actual service in March (the project to develop this R-5A missile had been codenamed N-1 in NII-88; the H-2 project was the development of more compact ballistic missile; the H-3 was the plan to develop two-stage intercontinental ballistic missile).[46] While the strategic importance of rockets was still not clear, the inaccuracy and expensiveness of V-2 was a reliable ground to make people believe that the interest in it was a mere transient fever.[47]

Even such leading Soviet engineers as Korolyov and Boris Chertok presented their objections to the policy of simple reproduction of "V-2." In 1947, Korolyov noted, "It would be erroneous to consider that the implementation of a domestic A-4 ('Vau-2' . . . *Ichikawa*) missile can be reduced to the task of copying German technology only with the replacement of materials with materials of domestic marks."[48] Chertok believed that "the A-4 rocket was morally obsolete in 1945."[49]

The above-mentioned on-the-spot inspection report by People's Commissariat of Aviation Industry[50] evaluated small anti-aircraft rocket guns developed in Germany rather than V-2. V-2 had a range of only 550–600 km and became available in actual service only in the end of 1944 due to lags in timing of explosion of warheads and inaccuracy of control, despite that the engine using alcohol and liquefied oxygen was already completed in 1942. This proves that this People's Commissariat did not expect much of V-2 rocket. This report stated, "Since these tasks have an aspect as an artillerist one, so the custody of equipment [obtained] in Peenmünde should be charged to People's Commissariat of Ammunition."[51]

How did the Ministry of Armament evaluate V-2? The resolution of the Collegium (Kollegia) of Ministry dated March 23, 1946, mentioned nine technological subjects concerning rocket development.[52] However, V-2 was out of their concern; there was no reference to it. Only in the second half of April, immediately after this resolution, did Ustinov begin to appeal to the top echelon of the political power to permit his Ministry to reproduce V-2. It seems that from the end of March to the first half of April, Ustinov and the leaders of the Ministry of Armament became aware of its significance and drastically changed their attitude toward V-2.

However, Ustinov reported the results of their research on V-2 to Stalin, Beria, and Malen'kov in a letter dated June 5, 1946. It referred straightforward to the technological incompleteness of V-2. The control devices of the Vau-2 missile by radio waves were cumbersome. More importantly, the missile lost control by the British radio interference. The autonomous locating system consisting of gyroscopes, the development of which had been sought for in Germany, was still in trial stage. The quantity of explosives that could be loaded in a warhead was only 800–1,000 kg. Although the flight range of the Vau-2 was 250–270 km, dispersion from the target was large, i.e., within ±10–20 km.[53] Nevertheless, this report approved of the acquisition of technology of V-2 rocket and recommended to start research and development activities for its reproduction.[54] A reporting memorandum for Stalin dated June 24, 1946, more concisely clarified the reasons.[55]

(1) It is possible for the Soviet Union to obtain the technology of V-2.
(2) If we could obtain V-2 technology, we should be given an opportunity to gain the fundamentals of newest reactive technology and more powerful reactive armaments.
(3) As (the development of) anti-aircraft rocket guns have just started, scientific research and design must also be conducted in the Soviet Union.
(4) The application of reactive armaments to battles is still predictive, and its character is not clear. Let the Ministry of Armed Forces (on February 25, 1946, the Ministry of Defense and Ministry of Naval Fleet were unified into this Ministry. Later, on February 25, 1950, it was divided into two Ministries; Ministry of War and Ministry of Naval Fleet. They were then integrated into the Ministry of Defense on March 15, 1953. – *Ichikawa*) investigate.
(5) It is required for the Soviet State to establish a new strong scientific research institute.

How did the military administration, to which the Ministry of Armament passed the question, evaluate V-1, V-2? The words V-1 and V-2 emerged on the journal, *Voennaya Mysl'* (*Military Thoughts*), edited by the Staff Academy for the first time on January–February 1945. However, the large, long-range rockets such as V-1 and V-2 were excluded from the consideration in this article, which was devoted to the domestic developments of short-range rockets and rocket guns.[56] German rockets, especially V-2, were thoroughly examined only in December 1946. The high effectiveness of reactive propulsive force was explained in comparison with explosive propellants like guns, and the situation on rocket developments in Western countries was introduced based upon the American and British journals.[57] The author of this article, General Pavel Kuleshov (1908–2000), says, "Rockets made rapid progress during World War II. However, it has only just begun." Kukeshov appeals for the rocket development for his fear that the Soviet Union might fall behind Western countries but not because of his conviction of prospectiveness of rocket technology.[58]

Generally speaking, it cannot be said that both sides of Ministry of Armament, which would take charge of rocket development, and the military leadership, which ought to have expected a promising feature in rocket technology, were not so firmly convinced of the strategic and tactical importance of rocket technology.

2.3 The start of rocket development by the Ministry of Armament

What was the reason the Ministry of Armament started the project from full copying of V-2 in spite of oppositions from Korolyov, Chertok, and other leading engineers?

M. Harrison assumed that the leadership of the Ministry of Armament rushed to the course to copy V-2 in full scale so as to make up for the technology gap between Germany and the Soviet Union.[59] Probably, the gap in the technical level between the Soviet Union and the Third Reich was very high. According to the report prepared by Korolyov,[60] of the total 1,271 plastic parts fabricated by pressing, casting, or rolling, only 1,012 items could be reproduced in the Soviet Union. Of them, 256 items were accomplished by manual operation using full-scale paper patterns. Of the 86 sorts of steel adopted in V-2, only 32 were produced in the Soviet Union. Of the 59 sorts of non-ferrous metals, only 21 items were provided in Soviet Union. In the factories affiliated to NII-88, the shortage of press machines, molds, and others was also serious. It ought to have been easily understood that the efforts for mere copying would include much loss since it was objectively required to develop rockets having more efficiencies than V-2 whose various shortages have already been pointed out. Therefore, this argument is not persuasive.

The presence of a large amount of idle plants and equipment, which were accompanied with the conversion of military production to peaceful production, should be noted as a more significant factor. Although the author could not find any document showing the net working rates of the affiliated plants, our examination here in the previous sections of this chapter reveals that the Ministry of Armament had a huge scale of idle production capacity. Chertok, who opposed the policy of mere copying V-2, was repeatedly persuaded by the leadership of the Ministry to understand the need for this policy. Later in his memoir, Chertok summarizes the persuasion in four points as follows:[61]

(1) Necessity to provide concrete and definite subjects in order to give a large number of workers the demands and the supports for their living and training.
(2) Impropriety to have various plants idle.
(3) State central testing ground was established during the war.
(4) Necessity for the motherland's industry to obtain rocket technology as soon as possible.

Ustinov became more positive for rocket development since he consented to the address of Gaidukov. A reporting memorandum addressed to Stalin, dated April 17, 1946, under the joint signature of Beriya, Malen'kov, Nikolai Bulganin (1895–1975), General Vannikov (1897–1962), Marshal Nikolai Yakoblev (1898–1962), and Ustinov,[62] which was prepared on Ustinov's initiative,[63] recommended to charge exclusively the Ministry of Armament with the tasks related to scientific research, design, and test research of guided and unguided liquid-fuel rockets and charge the Ministry of Agricultural Machine-building, which succeeded People's Commissariat of Ammunition, with the tasks related to solid fuel rockets and rocket guns. Gosplan instantly supported this proposal. In the above-mentioned letter to Beriya dated April 29, Gosplan recommended to order the Ministry of Armament to take charge of reproduction of V-2.[64]

Stalin accepted this recommendation.[65] On the basis of the proclamation of the Council of Ministers dated May 13, 1946, "Special Commission on Reactive Technology" affiliated to the Council was established with Malen'kov as Chair. At the same time, the Ministry of Armament took the main responsibility for the realization of the subjects as the executive organ. This commission consisted of members such as, except for Malen'kov, Ivan Zubovich (Deputy Chair), Ustinov, Yakovlev, Kirpichinikov, Aksel' Berg (1893–1979), Pyotr Goremykin (1902–1979), Ivan Serov (1905–1990), and Naum Nosovskii (1905–1978). Subsidiary subjects were assigned to various ministries. For the sake of realization of those subjects, new main directorates were organized in the Ministries of Agricultural Machine-building, Electric Machine-building, and Aviation Industry. The Ministries of Chemical Industry, Ship-building, and Machine and Instrument-building newly established their directorates. In June, the Ministry of Armed Forces established "the Fourth Scientific Research Institute (for Reactive Armament)" headed by General Aleksei Nesterenko (1908–1995).[66] Furthermore, the Ministry of Armed Forces established the "Directorate of Reactive Weapons" within the Main Directorate of Artillery and State Central Testing Site for Rockets. The Division of Rocket Technology was also established.[67]

The "State Commission for Survey of Experiences and Research on Rocket Development" was established in Germany in August 1946, and the basic subjects of "Nordhausen" Institute were confirmed. The activities in Germany became intensive, followed by the transfer of the research centers to the Soviet territory. The concentration of the activities in the domestic research centers was scheduled. In early August 1946, a governmental commission was held with Yakovlev as Chair in Breucherode. Ustinov, General Lev Gonor (1906–1969), Sergei Vetoshkin (1905–1991, a chief of a main directorate of Ministry of Armament), Georgii Pashkov (1909–1993, Chief, Section of Defense Industry of Gosplan), and others took part in this meeting. There, the framework of the Institute No. 88 was made up.[68]

All the Soviet persons were to move out of "Nordhausen" Institute and return to their homeland by the end of 1946. All the German engineers and workers who had been compelled to "corporate" with the Soviet administration were also taken to the Soviet territory in two suites of railroad carriages.[69] Each suite was composed of 68 carriages carrying some finished examples of V-2, the materials and the parts for 40 examples of V-2, and 380 German scientists and engineers. A comprehensive plan was settled to mobilize a variety of research institutes and design bureaus and allot the technological subjects to them so as to develop rockets within the Soviet territory. More than 40 research and design facilities extending over 12 ministries and one main directorate were mobilized.[70]

As mentioned before, the Ministry of Armament established Institute No. 88 on the basis of Factory No. 88 at Kaliningrad in Moscow Oblast. Its first Director was Gonor, Chief Designer – Pobedonostsev who had been in charge of the development of the so-called Kachusha rockets. Institute No. 88 consisted of four departments: Scientific Section, Special Design Bureau No. 88, Experimental Works No. 88, and Test Site. The Special Design Bureau No. 88 had some design offices; the third office was headed by Korolyov who was responsible for the reproduction of V-2, the designs of R-1 and R-2. In the fourth office, E.V. Sinel'shchikov and other designers engaged in the reproduction of *Wasserfall*, a long-range, guided, anti-aircraft rocket gun. In the fifth office, S.E. Rashkov and others were engaged in the reproduction of *Schmetterling* and *Reintochter*, middle-range, guided, anti-aircraft rocket guns. In the seventh office, Kostin and others were involved in the reproduction of *Taifun*, unguided anti-aircraft missile (both of liquid fuel and gunpowder propulsion). N.D. Umanskii and others from the Eighth Office took charge of the research and development of liquid-fuel engines with the development of oxidizer as a central issue. However, the team led by Aleksei Isaev (1908–1971) from the First Research Institute of Ministry of Aviation Industry succeeded in the development of oxidizer earlier than the Umanskii team. Therefore, the eighth office was closed in 1948 and absorbed into the Second Joint Design Bureau (another name – the ninth office of Institute No. 88) headed by Isaev.[71]

The Scientific Session was led by Chief Designer Pobedonostsev and consisted of some divisions such as Division M, whose chief was V.N. Iordanskii, that was responsible for material study. The other divisions were Division P (Chief – Viktor Panferov: 1916–1997) for strength study, Division A (Chief – Khalil Rakhmaturin: 1909–1988) for aerodynamics and hydrodynamics, Division I (Chief – Pavel Tsybin: 1905–1992) for a variety of testing methods, and Division U (Chief – Chertok) for controlling systems. The Chertok team carried out their research making full use of the precision machines for optical instrument works, which were spoiled as "trophies" in Germany. Later, they obtained the cooperation from the Institute of Automatic Machinery and Remote-controlling Devices of the Academy of Sciences.[72]

The reproduction of V-2 was soon completed, and the launching tests of R-1 (a copy of V-2) were carried out in 1948. A total of eight out of ten trials succeeded. In 1950, this type of rocket, which had a range of 270 km with a margin of error within 8 × 4 km, was deployed in the troops.[73]

3 Consequence of the financial problem of Ministry of Armament

What kind of effects did the reproduction of V-2 have on the serious financial condition of Ministry of Armament?

The ministry repeatedly asked the Ministry of Finance for additional financial aids, whereas the Ministry demanded to cover the deficit amounting to 150 million rubles in July 1946.[74] In 1947, it demanded compensation for the shortage of the revolving fund reaching 146 million and 600 thousand rubles (as of January 1), rescheduling of the settlement of the loans, supply of the credit amounting to 160 million rubles for the settlement in March,[75] additional supply of currency of 84 million rubles from the reserve fund of Council of Ministers in July,[76] compensation of deficit for 60 million rubles, and supply of the credit for the settlement for 160 million rubles in December.[77] In 1948, the Ministry requested compensation of the deficit amounting to 125 million rubles and the credit for the settlement of the loans in February,[78] additional supply from the reserve fund of Council of Ministers reaching to 154 million and 900 thousand rubles,[79] supply of 56 million rubles for the settlement of the debt and the loan secured exceptionally by the Industrial Bank for the purchase of the equipment in March[80] and rescheduling of the settlement of the loans to Industrial Bank in October.[81] Meanwhile, the proclamation of the Council of Ministers No. 3168–1036s dated September 9, 1947, lightened the planned target of the total output by the Ministry of Armament from 5,800 million rubles to 5,450 million rubles, increased the allotment of state subsidy and the budget for revolving funds from 242 million 90 thousand rubles to 339 million 840 thousand rubles and from 277 million 360 thousand rubles to 330 million 920 thousand rubles, respectively, and exempted the payment of 62 million 600 thousand rubles from the refund of profits to the state treasury.[82]

Thus, the Ministry of Armament could obtain huge amounts of additional financial support from the government. Although the relationship between those financial measures and the rocket development was not clarified directly in any document examined by the author, some documents of the Ministry of Finance revealed that the Ministry of Armament demanded financial aids for reasons related with the reproduction project of V-2. The Ministry of Armament requested additional aids of 20 million rubles for the employment of German specialists in October 1946.[83] It demanded a supply of 50 million rubles as capital investment for the establishment of Institute No. 88, including 19 million 700 thousand rubles for the arrangement of the railroad carriages for the transfer of materials and persons from Germany in December 1947.[84] In October 1948, the Ministries of Higher Education, of Armed Forces and of Electric Power Generation were obliged to pay from their own funds the sums (484 thousand rubles, 500 thousand rubles and 320 thousand rubles, respectively) needed for the research subjects with which Institute No. 88 charged the research facilities affiliated to those ministries.[85] In March 1949, the Ministry of Armament demanded payments of 3 million 200 thousand rubles for the construction of a jet-propelled engine testing station and an oxygen factory[86] and 4 million 319 thousand rubles for the purchase of carriages specialized for the tests of parts for "new technology," which Institute No. 88 ordered as an excess out of the plan.[87] In October, the Ministry demanded an increase of the allotment from the national budget for capital constructions concerning the projects of Institute No. 88 by 15 million rubles.[88]

The overall scale of financial aids related to the rocket development and its ratio in the total income of the Ministry are not clear. However, without any doubt, we can say that the Ministry of Armament seized some "sanctuary" to draw funds easily.

Notes

1 N.S. Simonov, *Voenno-Pronyshlenyi Kompleks SSSR v 1920–1950-e gody*. Moscow, ROSSPEN, 1996, 124.
2 Ibid., 125–132.
3 Ibid., 138.
4 Ibid., 139, 163.
5 Ibid., 139.
6 Ibid., 159.
7 Ibid., 138–140.
8 Ibid., 144.
9 Ibid., 140.
10 Ibid., 140, 141.
11 Yu.D. Maslyukov and E.S. Glubokov, "Planirovanie i finansirovanie voennoi promyshlennosti v SSSR." in A.V. Minaev (Ed.), *Sovetskaya Voennaya Moshch' ot Stalina do Gorbachova*. Moscow, Voennyi parad, 1999, 86, 105.
12 Simonov, *op. cit.*, in note 1. 153.
13 V.N. Krasnov, "Lend-Liz i razvitie Sovetskogo flota." *Voprosy Istorii Estestvoznaniya i Tekhniki*, No. 1, 1993, 69–80.
14 Simonov, *op. cit.*, in note 1. 185.
15 Ibid., 193.
16 Ibid., 217.
17 Ibid., 193, 194.
18 Ibid., 194, 195, 197.
19 Rossiiskii gosudarstvennyi arkhiv ekonomiki (Russian State Archive of Economy. Hereafter, RGAE). Fond (F.) 8157, Opis' (Op.) 1, No. 1047, 123.
20 RGAE, F. 8157, Op. 1, No. 2542, 54.
21 Ibid., 39–53.
22 RGAE, F. 8157, Op. 1, No. 2546, 7.
23 RGAE, F. 8157, Op. 1, No. 2572, 41. The State Optics Institute made great contributions in the Soviet history of science. An outstanding physicist specialized in optics, Dmitrii Rozhdestvenskii, and the future President of the Academy of Sciences, Sergei Vavilov, and other eminent physicists were enrolled. This institute played a crucial role in the introduction of new physics of the 20th century. For example, as early as in 1932, the news of the discovery of neutron was reported as a main topic in the seminar held in this institute for the first time in the Soviet Union. However, in the complicated ideological situation in the second half of 1930s, the institute was heavily criticized for its isolation from the practice (V.P. Vizgin, "Yadernyi shchit v 'Tridtsatiletnei voine' fizikov s nevezhestvennoi kritikoi sovremennykh fizicheskikh teorii." *Uspekhi Fizicheskikh Nauk*, Vol. 169, No. 12, December 1999, 1366, 1371). Eventually, its scientific branch was transferred to Leningrad State University in spite of the strong opposition from many physicists in the Academy of Sciences, including Rozhdestvenskii himself (*Sankt-Peterburgskii filial Arkhiva Rossiiskoi Akademii nauk* – SPb ARAN – F. 341, Op. 2, D. 22, 1, 17, 19–20, 83, 84, 99, 100, 119). Such a treatment brought serious negative influence on the Soviet physics. Nevertheless, some researchers of this institute like Mikhail El'yashevich took part in the Soviet nuclear project and made some important contributions (M.A. El'yashevich, "Podgotovka i provedenie opticheskikh nablyudenii pervogo Sovetskogo yadernogo bzryba." *Voprosy Istorii Estestvoznaniya i Tekhniki*, No. 2, 1996, 151–165).
24 RGAE, F. 8157, Op. 1, No. 2546, 6.
25 RGAE, F. 8157, Op. 1, No. 2545, 6.
26 RGAE, F. 8157, Op. 1, No. 1012, 78.
27 RGAE, F. 8157, Op. 1, No. 1047, 147.

28 3,653 persons or 1.4% of all employees in the Ministry in 1946. The ministry was constantly suffering from the shortage of manpower, especially of qualified workers; whereas 49,301 persons were newly enrolled in 1946, totally 70,138 persons left their jobs due to retirement for "domestic conditions," "illness," "going to schools or military service," "voluntary will," and death or transference to other ministries (RGAE, F. 8157, Op. 1, No. 2542, 59, 60).
29 For example, Factory No. 536 achieved 107.8% of the planned output. However, as for the military goods, only 39.8% was achieved and as for the goods for the civilian use – 89%. Such default was compensated by the excess of the production of goods demanded out of the plan, which extended to 555% (ibid., 59).
30 Ibid.
31 RGAE, F. 7733, Op. 36, D. 2110, 138.
32 RGAE, F. 7733, Op. 36, No. 2312, 102.
33 V.F. Utkin and Yu.A. Mozzhorin, "Raketnoe i kosmicheskoe vooruzhenie." Minaev (Ed.), *op. cit.*, in note 11, 208–210. Tukhachevskii made a great effort for the sake of the establishment of RNII. He also had such foresight that he was interested in the use of electro-magnetic method for antiaircraft defense and ordered the selection of suitable research facilities (B.V. Sokolov, *Mikhail Tukhachevskii: Zhizn' i Smert' "Krasnogo Marshala."* Smolensk, Rusich, 1999, 318).
34 L. Beriya, G. Malen'kov, N. Bulganin, B. Vannikov, D. Ustinov and N. Yakovlev, "Dokladnaya zapiska I.V. Stalinu (17 aprelya, 1946g); Ob organizatsii nauchno-issledovatel'sikh rabot v oblasti raketnogo vooruzheniya v SSSR."/AP RF, F. 3, Op. 47, D. 179/(publikatsiya V.I. Ivkina), *Voenno-Istoricheskii Zhurnal*, No. 3, 1997, 33.
35 S.P. Korolyov, "Dokladnaya zapiska; K voprosu o rabotakh Byuro samoletnykh reaktivnykh ustanovok pri OKB RD na zavode No. 16." 30 sent. 1944g/AMMK/, Pod obshch. Red. B.V. Raushenbakha, *S.P. Korolyov i Ego Delo: Svet i Teni v Istorii Kosmonavtiki.* Moscow, Nauka, 1998, 83.
36 Simonov, *op. cit.*, 231.
37 S. Averkov, "Rakety Tret'ego Reikha (interv'yu s V.S. Budnikom)." *Rabochaya Tribuna*, 4 iyulya 1991, 32, 33.
38 A.I. Shakhurin, "Dokladnaya zapiska G.M. Malenkovu ot 8 iyunya 1945g."/AP RF, F. 3, Op. 47, D. 182/(publikatsiya V.I. Ivkina), *Voenno-Istoricheskii Zhurnal*, No. 3, 1997, 32, 33.
39 B.E. Certok, *Rakety i lyudi* (2-e izdanie). Moscow, Mashinostroenie, 1999, 195, 196.
40 Beriya i dr., "Dokladnaya zapiska I.V. Stalinu . . ." *op. cit.*, in note 34. 33.
41 Utkin and Mozzhorin, *op. cit.*, in note 33. 210.
42 RGAE, F. 8157, Op. 1, No. 1073, 14. According to a report dated June 1, 1946, Ministry of Armament had obtained 981 detailed drafts of V-2 out of totally 2,936 drafts, 75% of the V-2 rocket engine parts, and one actual set of the automatic controlling devices for V-2, which was used in the air raid over London and was so simple and already outdated (ibid., 136).
43 Ibid., 18–20.
44 Ibid., 22–24.
45 Ibid., 46.
46 Utkin and Mozzhorin, *op. cit.*, in note 33. 216.
47 M. Harrison, "New Postwar Branches (1): Rocketry." in J. Barber and M. Harrison (Ed.), *The Soviet Defense-Industry Complex from Stalin to Khrushchev.* London, Macmillan Press, 2000, 126.
48 S.P. Korolyov, "Zametki po raketnoi tekhnike (1947g)."/APPK/, Pod obshch. red. Raushenbakha, *op. cit.*, 114.

49 B.E. Chertok, *op.cit.*
50 Shakhurin, "Dokladnaya zapiska . . ." *op. cit.*, in note 38. 32, 33.
51 Ibid., 33. Shakhurin and his ministry took charge of jet airplane development, which was a huge, complicated scientific and technological project as described in the next chapter. It seems that Shakhurin's negative attitude toward rocket development was brought about by this reason.
52 RGAE, F. 8157, Op. 1, No. 1046, 7. The technological subjects cited that there were (1) guided antiaircraft rocket guns; (2) unguided antiaircraft rocket guns; (3) automatic homing antiaircraft rocket guns (long-range and short range); (4) application of reactive technology to antitank artillery; (5) development and improvement of antiaircraft rocket engines; (6) stabilizing, radio control, automatic control, and homing devices for rocket guns (utilizing photo elements and other types); (7) launch pads; (8) development of testing methods of rocket guns; and (9) hydrodynamic problems concerning navigation of rockets at the highest speed.
53 Ibid., 66–69.
54 Ibid., 70, 71.
55 D. Ustinov, I. Kabanov, N. Yakovlev, I. Serov, P. Goremykin and N. Nosovskii, "Dokladnaya zapiska I.V. Stalinu ot 24 iyunya 1946g."/AP RF, F. 3, Op. 47, D. 182/(publikatsiya V.I. Ivkina), *Voenno-istoricheskii zhurnal*, No. 3, 1997, 37.
56 V. Khokhlov, "Razvitie artillerii vo Vtoroi mirovoi voine." *Voennaya Mysl'*. yan.-feb. 1945. 28.
57 P. Kuleshov, "Reaktivnaya artilleriya i perspektivy ee razvitiya." *Voennaya Mysl'*, dek. 1946, 45–50.
58 Ibid., 50.
59 M. Harrison, "New Postwar Branches (1): Rocketry." in J. Barber and M. Harrison (Ed.), *The Soviet Defense-Industry Complex from Stalin to Khrushchev*. London, Macmillan Press, 2000, 126.
60 Korolyov, "Zametki . . ." *op. cit.*, in note 48. 112–114.
61 Chertok, *op. cit.*, in note 49. 306.
62 Beriya i dr., "Dokladnaya zapiska I.V. Stalinu . . ." *op. cit.*, in note 34. 34.
63 Ya.K. Golovanov, *Korolyov: Fakty i Mify*. Moscow, Nauka, 1994, 362–364.
64 RGAE, F. 8157, Op. 1, No. 1073, 14.
65 Simonov, *op. cit.*, in note 1. 233.
66 V.I. Ivkin (publikatsiya S. Kharlamova), "Cherez ternii d kocmos." *Armeiskii sbornik*, No. 6, 1994, 74.
67 Utkin and Mozzhorin, *op. cit.*, in note 33. 211.
68 Chertok, *op. cit.*, in note 49. 171–174.
69 Utkin and Mozzhorin, *op. cit.*, in note 33. 210. As a result of observation on the German engineers for half a year, it became clear that the Germans could not build up rockets for themselves. Consequently, approximately 150 German engineers (about 500 in all with their families) including 13 professors, 32 doctors, and 85 diplomates were transferred to the institute No. 88 as "the first branch" of the institute (headed by F.G. Sukhomlinov) and were compelled to be isolated. In spite of such a treatment, Helmut Gröttrup, a leader of the German engineers, tried to maintain their own unity by appealing to the administration of the institute a plan that they, the Germans, took responsibility to build a new rocket, G-1, for themselves. However, this proposal was shelved in September 1947, as not an urgent subject. During the internment, the German specialists were isolated in Gorodmul' island and were only sometimes asked technological questions because of confidentiality and the opposition from the side of Soviet engineers such as Korolyov, who were so proud of their own ability that they desired to achieve the subjects for themselves (Chertok, *op. cit.*, in note 49. 198, 201–207).

70 Malen'kov, Bulganin, Ustinov, Khrunichev, Yakovlev, Pavlenko and Kirpichnikov, "Dokladnaya zapiska I.V. Stalinu ot 31 dekabrya 1946 goda."/AP RF, F. 3, Op. 47, D. 183/(publikatsiya V.I. Ivkina), *Voenno-Istoricheskii Zhurnal*, No. 3, 1997, 38, 39.
71 Chertok, *op. cit.*, in note 49. 247; Utkin and Mozzhorin, *op. cit.*, in note 33. 212. After the Leningrad Doctors' Plot Affair that broke out in 1948, the Jewish people in the Soviet Union became repressed on 'Zionism' or 'cosmopolitanism.' The arrest of Gonor in 1949 brought about a certain difficulty to the institute (Chertok, *op. cit.*, in note 49. 199, 245.) Yakovlev and Shakhurin were also arrested (later after the death of Stalin, their honors were restored. V primechaniyakh V.I. Ivkinym, *Voenno-Istoricheskii Zhurnal*, No. 3, 1997, 40, 41).
72 Chertok, *op. cit.*, in note 49. 247.
73 Utkin and Mozzhorin, *op. cit.*, in note 33. 214.
74 RGAE, F. 7733, Op. 36, No. 2110, 140.
75 RGAE, F. 7733, Op. 36, No. 2302, 59.
76 Ibid., 153.
77 RGAE, F. 7733, Op. 36, No. 2571, 26.
78 Ibid., 110.
79 Ibid., 224.
80 Ibid., 184.
81 RGAE, F. 7733, Op. 36, No. 2572, 160. As of January 1, 1948, the total sum of the debts owed by the Ministry of Armament amounted to 349 million 300 thousand rubles and the shortage of revolving funds – 125 million rubles. Ustinov accounted for such a great deal of deficit to the uncollected credits bills owed by other ministries to the Ministry of Armament, which reached 516 million 700 thousand rubles (RGAE, F. 7733, Op. 36, No. 2571, 111). However, it must be true that Ustinov used a far-fetched argument. Generally speaking, there is no guarantee that the possession of a large amount of credits and bills will cover the debts of the creditor. The financial problem of the Ministry of Armament was so deep-rooted that it did not grant any optimistic view.
82 RGAE, F. 7733, Op. 36, No. 2303, 62. Among the documents of the Ministry of Finance, there is one file titled "Conclusion on the issues commissioned by the Council of Ministers concerning the financial problems of Ministry of Armament," which is not allowed to be viewed ("Zaklyuchenie po porucheniyu Sovmina SSSR po voprosu finansirovaniya Ministerstva vooruzheniya SSSR, 10 fev.–21 iyulya 1948g."/RGAE, F. 7733, Op. 36, No. 2573). After the issue of this document in the summer of 1948, few documents showed the request from the Ministry of Armament for financial aids (except for the documents concerning the institute No. 88). It is possible that some bold measure such as exemption of the debts might have been taken.
83 RGAE, F. 7733, Op. 36, No. 2110, 187.
84 Ibid., 140.
85 RGAE, F. 7733, Op. 36, No. 2572, 166.
86 RGAE, F. 7733, Op. 36, No. 3020, 55.
87 Ibid., 50.
88 RGAE, F. 7733, Op. 36, No. 3021, 50.

4 Plunder of technology

Jet plane development and German technology

The development of jet planes in the Soviet Union was a process of a huge scale of "plunder of technology" from a country defeated in the war orchestrated by a victorious country on an unprecedented scale. However, despite the attempt to copy the German jet planes, the early Soviet jet planes were not simple copies of German airplanes. What were the reasons behind this outcome? To what extent did the Soviet Union succeed at learning German technology, and how limited was it?

1 The domestic development before the end of World War II

1.1 Development of turbo jet engine

In 1929, Boris Stechkin (1891–1969), chief of motor engineering unit of Central Aero-hydrodynamic Institute (*Tsentral'nyi Aerogidrodinamicheskii Institute*; TsAGI), published his thesis "Theory of Aircraft Reaction Engine" in the magazine *Tekhnika Vozdushnogo Flota* (*Technology of Airfleet*).[1] Subsequently, in the mid-1930s, a student of the Khar'kov Aviation Institute named Anatolii Eryomenko conceived the first turbo jet plane. It was named KhAI-2 after the initials of the Khar'kov Aviation Institute in Russian. An engineer of the heat engine laboratory of the same school, Arkhip Lyul'ka (1908–1984), designed its engine RTD-1 (traction of 500 kg).[2] The design process is said to have started in 1937.[3] KhAI-2 was designed to have a maximum speed of 500 km/h, regular speed of 120–130 km/h, and weight of 1.5 t, however, now KhAI-2, also RTD-1, only remain on paper. Lyul'ka also engaged in the production of a new engine, RD-1 (the initials of the Russian words, *Reaktivnaya dvigatel*; Reactive Engine) at Kirov Factory in Leningrad and 75% of the design work was completed in the summer of 1941.

In October 1942, Mikhail Gudkov (1904–1983), who was developing a fighter aircraft along with Semyon Lavochkin (1900–1960) and Vladimir Gorbunov (1903–1945), came across the drawings of RD-1 at the People's Commissariat of Aviation Industry and conceived the LaGG-3 (a combination of the initials of the three designers), equipped with RD-1. The designed capacity of RD-1 included traction of 530 kg and maximum speed of 700 km/h. However, according to Tursun Sultanov, who studies the history of jet plane development in the Soviet Union, this estimation on planned performance was overoptimistic and unrealistic.[4] Therefore, it required deeper analyses. However, as the war intensified, Lyul'ka was unable to develop his studies under the so-called Leningrad Blockade. In 1943, the government issued a decree entrusting an experimental factory to Lyul'ka,[5] but he would return to his study only in 1944 – even after he escaped from Leningrad, and he fell sick.[6]

Before being able to use the engines acquired from Germany as "war trophies," the Soviet Union had to use the engines developed by themselves. In early 1944, the MiG team (a combination of the first letters of the surnames of the two designers), led by Artyom Mikoyan (1905–1970, the younger brother of Anastas, a powerful politician) and Mikahil Gurevich (1892/93–1976), and the Pavel Sukhoi (1895–1975) team were, respectively, ordered to develop the I-300 [*Istrebitel* (fighter aircraft)-300] and Sukhoi-9 (abbreviated as Su-9). Although the result was a failure owing to the incompleteness of Lyul'ka's engine,[7] the design of I-300, its control system and equipment were to be installed in the first jet plane of the Soviet Union, MiG-9 (to be mentioned later), as they were.

1.2 Development of other reactive engines

The maximum speed of military aircrafts mass-produced in various countries increased by approximately 100 km/h in the first four years of the war, from 500–550 km/h to 600–650 km/h. The horsepower of aircraft engines almost doubled from approximately 1,000–2,000 hp. However, the Soviet Union was lagging in terms of both power and maximum speed. Soviet aircraft engineers began their attempts to equip a DC-type ramjet engine (simple type of jet engine composed of an air intake, combustor, and nozzle, without compressor) to a fighter aircraft in 1940. The fighters, I-15bis and I-153, were both installed with DC-type ramjet engines under the wings in addition to the propeller engines, which were the main propulsion forces. Such engines were also installed in LaGG-3 and Yak-7B (an aircraft developed by the design bureau of Aleksandr Yakovlev, abbreviated as Yak) as an experiment. Although this increased the speed by approximately 30–50 km/h, the speed decreased when the ramjet engine was inactive and while it was active, the fuel consumption reached such a high rate of 60–70 kg/min.[8]

The Soviet engineers subsequently sought to develop another type of engine that combined a regular propeller engine and some basic elements of jet engine which was still underdeveloped at that time by means of synchronized operation of a propeller engine and a jet engine with a single axis compressor attached (the 900 hp "Isotta-Fraschini" turbo prop engine, developed by the Italian company Caproni in August 1940, was the first of such trials). Between 1943 and 1945, a team led by Konstantin Kholshchevnikov (1906–1976) engaged in the development of this type of engine; through March and April 1945, they equipped engine models I-107 and I-250 on a Su-5 and a MiG-13, respectively, and conducted test flights. However, due to the large weight of the piston engine and the low efficiency of propeller engine during high-speed flights, the research and development of turbo prop engine was aborted.[9]

There were attempts to attach an auxiliary liquid-fuel rocket engine on Pe-2 (an aircraft developed by the design bureau led by Vladimir Petryakov – 1891–1942 – abbreviated as Pe) bomber, Yak-3, La-7 (an aircraft developed by the design bureau led by Lavochikin, abbreviated as La), and Su-7 fighter aircraft (with an RD-1 engine with a traction of 300 kg). Although the speed increased by 70–180 km/h, due to the necessity of storing liquid nitric acid as oxidizer, high fuel consumption of approximately 90 kg/min, unreliability of rocket engine, and difficulty in control, the results were far from satisfactory.[10]

Having learned that Germany had successfully developed a rocket plane, and that even Japan was working on its development, Viktor Bolkhovitinov (1899–1970) advised the government to develop a fighter aircraft equipped only with a single liquid-fuel rocket engine. Thus, it was decided that the research and development of the rocket plane, BI, would be carried out by Aleksandr Bereznyak (1912–1974) and Aleksei Isaev (1908–1971).[11] BI was equipped with a liquid-fuel rocket engine D-1A-1100 having traction of 1,100 kg on its airframe attached with regular-shaped wooden wings. Considering the serious situation in the war, BI was equipped with two normal 20 mm cannons from the initial stage of its research and development activities. However, due to the evacuation of factories to the east, the production was stalled and the first test flight was carried out on May 15, 1942. Despite lasting only three minutes, it became the first record of rocket flight. Experimental studies continued thereafter, but due to discoveries of dozens of defects soon after, the poor efficiency of achieving only two minutes of flight despite the load of 705 kg of fuel and oxidizer and finally an accident that occurred on March 27, 1943, the developmental studies were suspended.[12]

2 The research and development activities in occupied Germany

According to Yakovlev,[13] the Deputy of People's Commissar of Aviation Industry of that time, Stalin refused a request by Yakovlev who proposed in an assembly held at the Kremlin in February 1945 to follow the path of independent development and ordered the acquisition of German technology and use of engines manufactured in Germany. He prioritized the acquisition of German technology in the shortest period possible.

In March 1946, a "Special Commission" of the government chaired by Yakovlev made an inspection trip to the major aircraft factories in Germany. On April 2, 1946, Yakovlev and the Minister of Aviation Industry, Mikhail Khrunichev (1901–1961), requested Stalin to organize the investigation on the future image of aircraft engines in the Soviet Union. On the same occasion, Yakovlev presented a report on the inspection trip to Germany.[14]

The Yakovlev Commission's report was analyzed by the Council of Ministers, which on April 17, 1946, issued a resolution (Council of Ministers Proclamation

No. 874–366ss.) ordering experimental studies on the following ten topics in Germany: (1) production of Jumo-004F engine (traction of 1,200 kg); (2) production of Jumo-012 engine (traction of 3,000 kg); (3) production of BMW-003S engine (traction of 1,050 kg); (4) production of BMW-018 engine (traction of 3,400 kg); (5) production of low-altitude bomber, Ju-126 (maximum speed of 780 km/h); (6) production of bomber, Ju-131 (maximum speed of 860 km/h); (7) production of experimental plane, Siebel346 (maximum speed of 950 km/h); (8) design of high-speed jet bomber, Ju-132 (maximum speed of 950 km/h); (9) design of a powerful aircraft diesel engine Jumo-224; and (10) design of automatic piloting devices which function even without a pilot.[15]

2.1 1st Special Design Bureau, OKB[16]-1, Dessau

Special Design Bureau (*Ocoboe Konstryktorskoe Byuro.* Hereafter, OKB) No. 1 was established based on a Junkers factory in Dessau. As on May 14, 1946, the engine design office led by Dr. Scheibe (first name unknown) had 402 members (ten PhD engineers, 22 engineers with a diploma, 47 engineers, 90 technicians and designers, 88 draftsmen, eight supervisors – Meisters, 166 workers and staff members); the aircraft design office led by Brunolf Baade (1904–1969) had 433 members (two PhD, 31 engineers with a diploma, 164 engineers, 106 designers, 33 draftsmen, 95 workers, staff members, and two pilots). There were 254 members in the engine works, 635 in the airplane construction works and 191 in the common part works – totaling 1,080 (among them, 47 were engineers). Furthermore, the factory's engineering division had 66 engineers, 106 workers and staff members, and 851 personnel allocated in auxiliary departments such as a service division, garages, finance, accounting, personnel, and security. The entire OKB-1 had a total of 2,992 members.[17]

The following is an example of the "collaboration" of German specialists: Ferdinand Brandner (1903–1986) was the manager of Muldenstein Factory near Leipzig. This factory had started the production of Jumo-004 engine in December 1944. The manager was caught by Soviet troops and forced to collaborate to reassemble the production team members, who were scattered around in a great turmoil. He was thereafter taken to Moscow, introduced to Vladimir Klimov (1892–1962), and forced to report on the situation of jet plane development by Junkers in a conference. Subsequently, he was forced to work under Nikolai Kuznetsov (1911–1995), Klimov's adjutant, in Leningrad. Albrecht says that the Soviet Union was fond of people like him, who were in the third or fourth level of the hierarchy of a company. As for Brandner, he was released in 1953.[18]

The main tasks undertaken during this period were the development of the jet engine Jumo-004F, turbo prop engine Jumo-012, and aircraft diesel engine Jumo-224; the design and production of jet planes Ju-126, Ju-131, and Ju-132; and R&D of rocket planes. Here, we shall take a view of the reports issued during a very brief period before the transfer of R&D activities to the Soviet Union in the fall of 1946.

[Jumo-004F]

May 8. The design of Jumo-004F was modified: (1) the air intake was converted into a doubled structure; (2) the fitting of a stabilizing fin to the jet stabilizer was more stabilized; (3) the sheltering structure of the gas compressor was doubled owing to the improvements in the cooling device; (4) the tip of the jet nozzle was improved for the same reason; (5) the jet nozzle acquired a mechanism changeable in response to the quantity of intake air. These were because of a variety of malfunctions due to overheating in the jet nozzle.[19] Determining an adequate heat resistance of the jet plane was of great significance, as will be mentioned later.

By July 25, 1946, as for Jumo-004F, a total of 25 h of operational experiment and performance measurement tests had been already completed with the collaboration of Dr. Scheibe and his team.[20]

[Jumo-012]

The assembly of the first unit of Jumo-012 was completed on June 25, 1946, and was tested for the first time on July 6, 1946. However, due to the fragile structure, the rings inserted to the axis between impellers of the turbine was often damaged, the inner wall of the gas condenser distorted. Additionally, shrinking and distortion in the diameter of a guiding part of the turbine were caused by a design error.[21]

[Jumo-224]

The development of aircraft diesel engines for Jumo-224 was led by Dr. Scheibe. Later, he organized an independent group headed by an engineer Gerlach (first name – unknown), including 29 engineers and design engineers, ten technicians, 12 draftsmen, four copyists, totaling 55 members and made them dedicate exclusively to the development of diesel engines.[22]

[Jet airplane, Ju-126]

By May 10, 1946, a trial unit of Ju-126 (or, EF-126) for the flight tests had been completed, examined in ground tests and, after them, finished. The manufacture of the second unit had finished, and the third to fifth units were then under production.[23]

In June, the second unit of Ju-126 was completed and the towing test was successful. However, the following defects were detected: (1) the retraction of the ski-type taxiing device used for take-off and landing was too slow, (2) the receding angle of the elevator of the plane was insufficient, (3) the pilot lamp was often distorted, and (4) the differential gear of the elevator was too small. The third and fourth units were planned to be prepared in Soviet territory, and the possible changes were planned to be implemented in the fifth unit.[24]

[Jet airplane, Ju-131]

As of May 10, the airframe design for Ju-131 (or EF-131) had been completed; however, for the production of the first unit, the following items were still not ready[25]: (1) aiming device for shooting and bombing, (2) 12 starters with rocket propulsion, (3) autopilot and radar, and (4) individual control devices and electric measurement devices.

In June, a "Model Production Commission" was established, and the design of Ju-131 was already in its final stage. However, during the same period, the air force submitted a "Tactical and Technical Demands" report to the Ministry of Aviation Industry,[26] demanding a large upgrade of performances of Ju-131. OKB-1 responded that, "The model 131 that will appear at the airfield in September will be beneficial to the air force as is. We believe that it will serve both study and training purposes and offer the experience of piloting at high speed and high altitude, handling fighting equipment and jet bomber piloting," and rejected the demands.[27]

[Jet airplane, Ju-132]

Dr. Baade reported it to the Model Production Commission that as of July 25, all the materials related to Ju-132 had been already collected. However, the work in some of the Design Bureau sections was stagnant because some essential steel and non-ferrous metals were not standardized yet in the Soviet Union.[28]

[Yuliya Project]

The Yuliya (Julia) Project was aimed at developing aircrafts equipped with rocket engines. They were designed to reach high altitudes at high speed such that they could raid bombers. They required catapult take-off.[29]

Based on the proclamation No. 606–249s. issued by the Council of People's Commissars dated March 10 and the order No. 1/195 issued by Deputy Commissar, a specialized bureau was established in July with 149 specialists from the company Henkel[30] for catapult design.

2.2 2nd Special Design Bureau, OKB-2, Stassfurt

This bureau was established based on a BMW factory in Stassfurt. Initially, it had as few as 158 members in total – five staff members and 95 workers (among them, 60 were direct employees) in the works, 28 staff members and five workers in the design office, five staff members and 20 workers in the management – but by December 31, 1945, the total was boosted to 558, with 65 staff members and 263 workers in the works, 78 staff members and 22 workers in the design office, and 40 staff members and 90 workers in the management.[31]

As of May 27, 1946, the total number of employees was 1,723, among whom 308 were allocated to the research and development sections and 490 to the production lines. The number of Russian specialists was as few as 12.[32]

As of July 17, the total number of employees was 2,107 with 392 in the design office, 577 manufacturing workers, 157 in the intermediate and tools works, 400 auxiliary workers, 304 in the construction sites, 87 in the management, and 190 in auxiliary departments such as drivers, security, and firemen. There were 11 Russian specialists and 342 units of equipment.[33]

The following is a summary of the main tasks of this bureau: the developments of the two jet engines, BMW-003 and BMW-018.

[BMW-003]

As there were a few units of BMW-003 engine with the traction of 800 kg left, two test-beds – a starting motor test-bed and jet nozzle test-bed – were made before the end of 1945. The data collected from the operation tests often repeated were used for an improvement research aimed at obtaining traction of 1,050 kg.[34]

In the first burning test with a BMW-003 engine conducted in early July 1946, only 1 h and 5 min after the start three turbine wings were torn into pieces due to the high heat (850°C). The third unit, completed on July 15, featured some design changes. The burning test was carried out carefully in order not to exceed the regime of the speed of sound and traction of 800 kg. The temperature of the turbine wing measured using an optical pyrometer was 750°C. Even 1 h and 50 min after the start of the burning test, no defect was observed with only one exception of a very minor rattle in a single portion of the turbine wing.[35]

The first unit of the improved version of BMW-003S was assembled on June 15 and subjected to an operation test for 1 h and 39 min. A deformation in the central turbine wing was detected when one hour passed since the start of the test. The turbine wing of the second unit also broke when 40 minutes passed since the start. A large vibration also appeared in the turbine wings and overheating in the combustion chamber caused the turbine diameter to expand. It was reported that, due to such expansion, the distance between the turbine wing edge and the internal wall was narrowed from 2.8 mm to a dangerous level of 0.8 mm.[36]

[BMW-018]

All the documents for the BMW-018 engine had to be newly created.[37]

As of May 27, 1946, only 15% of the detailed drawings required for the production of the BMW-018 engine was completed with approximately 300 sheets and the production of parts also progressed to 10%. However, the report suggests that BMW-003 was prioritized over BMV-018. It also describes the difficulties faced by the engineers due to the lack of a cupola furnace for the thermal process.[38]

As of July 17, restoration of the detailed drawings of the BMW-018 engine progressed to 90% and the preparation of the parts had reached 25% of completion.[39]

The work plan progressed quickly, and as of July 25, it was 90% complete.[40]

2.3 3rd Special Design Bureau, OKB-3, Halle

This Bureau was established based on a "Siebel" factory in Halle. As of May 1, 1946, it had 788 employees and among them 131 were engineers.[41]

According to a report dated September 31, the subjects of the engine development section, which was one of the core sections of OKB-3 and whose final target was to develop a supersonic plane, were the developments of : (1) Walter 109–509 (traction of 1,700kg); (2) Walter 109–510 (traction of 2,000 kg); (3) BMW-3395 (auxiliary rocket engine for jet engine BMV-003. traction of 1,200 kg); (4) BMW-3390S (also an auxiliary rocket, traction of 4,000 kg); (5) eight test stations for Walter engine, BMW engine, pumps and turbines, and the study of the burning process; (6) the assembly of fuel tanks; and (7) fuel for liquid-fuel rocket related subjects.[42]

However, it appears that OKB-3 was facing various difficulties. The most serious one, the problem of securing fuel, will be mentioned later. According to the report dated May 17, the other difficulties they were facing were: (1) lack of leadership owing to the absence of Soviet specialists; (2) difficulty of securing German specialists (highly skilled German specialists like Singelman [accurate spelling and details – unknown] was caught and sent to the USA and even those who were left in areas occupied by the Soviet troops would not dare to work for the Soviet Union); (3) difficulty of acquiring raw material and equipment (the decision to deploy a transportation corps at OKB-3 was stalled); and (4) bureaucratic challenges (even if raw material and equipment were acquired, obtaining authorization to transport them was so time-consuming owing to the bureaucratic system).[43]

Most of the test stations, fuel-tank-related facilities, and raw material were in Basdorf as of May 17. According to a report, they were being gradually transported to OKB-3 in Halle. However, some of them were directly sent to Moscow, which aggravated the situation.[44]

Moreover, the shortage of the human resources was serious. Since its initiation, OKB was short on design engineers specialized in designs of engines and high-altitude cabins and chemists specialized in a special type of jet fuel.[45] Further, the human resources that could be secured in the areas occupied by the Soviet Union were limited. In a report dated May 17, a group of officers of OKB-3 who were in charge of selection of specialists stated that a large number of specialists lived in the areas occupied by the other Allies and, therefore, "prompt support from the appropriate institutions is necessary." Some German specialists were transported from the areas occupied by other Allies to OKB-3 via different methods. For example, some of the specialists involved in the development of Siebel 346 were transported from such areas by a general named Serov (first name and patronymic – unknown) in certain ways.[46] However, the outflow of valuable personnel was a more serious problem.[47]

The following is an overview of two of the main subjects: the development of the supersonic plane Siebel 346 and the engine Walter that was planned to be equipped to that plane and the problem of securing fuel.

[Walter engine]

When OKB-3 was established, in addition to a stock of semi-finished products delivered from Margulis, some auxiliary parts and units had already been gathered from the airfields and stockyards in Eilenburg and Brandiff. It was estimated that, if they had been repaired and complemented with a few missing parts, it would have been possible to build 14 Walter engines using them.[48]

The four variants of Walter 109–509 were all acquired as completed models. In particular, two units of A2 model were assembled and each tested on July 1 and 3. However, Walter 109–510 did not undergo a necessary burning test due to the lack of T-Stoff (liquid hydrogen peroxide) and S-Stoff (fuel element). Although three units of Walter 109–510 for Siebel-346 were completed so far, the burning test was not carried out due to the lack of fuel.[49]

[Siebel 346]

As of July 25, the experimental unit of supersonic plane Siebel-346 was still under construction. The design office engaged in the design of a glider for training, the purpose of which was to improve the maneuverability and safety of the airplane.[50]

[Fuel problem]

On May 17, 1946, a stockyard near Burg (Parey) was ordered to supply 150 t of hydrogen peroxide (H_2O_2). S-Stoff, the fuel element used by German air force for their jet planes, was discovered in the Brandiff airfield. However, due to obstruction by the garrison headquarter there, it could not be carried out.[51]

Walter required 82% concentrated hydrogen peroxide and S-Stoff as fuel, and the BMW engine required 98% concentrated nitric acid and a special fuel. The acquired 25 t of S-Stoff contained too much moisture, and a Buna (butadiene sodium rubber) factory was ordered to distill it. However, it was estimated that the quantity would not be sufficient, even though the acquired 30 t of fuel for the BMW engine contained excessive methyl alcohol. As S-Stoff and the fuel for BMW could not be produced in the areas occupied by the Soviet Union, as an alternative, it was suggested to purchase raw materials such as hydrated hydrazine, xylidine, triethyl amine, and others from the enterprise I.G. located in an area occupied by other Allies. Simultaneously, the development of substitutional methods using alternative raw materials obtainable in the Soviet occupied areas was being pursued – for example, phenyl-hydrazine instead of hydrated hydrazine and dimethyl isobutyl amine instead of triethyl amine.[52]

2.4 4th Special Design Bureau, OKB-4, Berlin

This Design Bureau, established on the basis of a factory of Askania Werke in Berlin, mainly focused on the study and development of autopilot system. As of March 25, 1946, there were working 56 employees. Specialists working at other companies were also being brought in. As of March 25, 1946, there were only three Soviet engineers: Director Leont'ev (first name and patronymic unknown), Deputy Director Sokolov (first name and patronymic unknown), and their Assistant, captain Reznikov (first name and patronymic unknown).[53]

OKB-4 established a design office (Figure 4.1), experimental works and laboratory (the First Works had an area of 1,500 km^2 with 45 machine tools, Figure 4.2) in Berlin and Friedrichshafen before July 1, 1946. The Second Works of area 3,000 km^2 with 48 machine tools was established soon after. As of July 1, the Bureau had four PhDs, six engineers with a diploma, 33 engineers, 16 operators, 39 staff members, 143 licensed workers, and 129 auxiliary workers – totaling 367 members.[54]

As of March 25, 1946, the Design Bureau had created approximately 350 sheets of detailed drawings. The targets of the Bureau were to develop the following items: (1) a central gyroscope station (a gyroscope mechanism equipped with a Ferraris motor, a gyroscope for control, and an electronic amplifier); (2) a controlling device; (3) an automatic signal device; (4) an electric power generator driven by a motor with 30 V and 500 Hz to supply three-phase AC current; (5) three types of piloting devices; and (6) a radio altimeter. Although some detailed drawings of the gyroscope mechanisms were preserved, most of the documents either were lost or did not exist as the company Askania Werke had not started

Figure 4.1 Design office of the 4th Special Design Bureau, OKB-4.

Source: Rossiiskii gosudarstvennyi arkhiv ekonomiki (Russian State Archive of Economy. Hereafter, RGAE). Fond (F.) 8044 Opis' (Op.) 1 No. 6676.50.

its design or did not create any documents then. Furthermore, the model of "Mutter-compass," which was one of the two compasses that would serve as a part of the gyroscope for control and "probe-type compass," was transferred to Moscow by Lieutenant Senilov (first name and patronymic unknown). Thus, initially, there were very few documents that could be used as reference.[55]

According to a report from July 25, 1946, there were 15 specific tasks of OKB-4 aimed at development and manufacture of: (1) a gyroscope station, (2) induction compass indicator, (3) central compass route indicator, (4) central piloting device (indicators and others), (5) balanced route indicator, (6) signal device, (7) control stick, (8) gearbox, (9) gliding angle adjustment device, (10) radio absolute altimeter, (11) V-shaped wireless receptor, (12) radio distance meter, (13) automatic calculation and decision device for take-off and landing, (14) automatic propeller regulator, and (15) transformer to send 500 Hz AC current to automatic piloting equipment.[56] OKB-4 was also responsible for the development of equipment for experimental studies, such as (1) a spindle for aerodynamics study for Mach indicator, (2) an indicator measuring the forces involved in the control of the airplane by control stick, (3) an indicator measuring the forces burdened on the coupling of a wing to the airframe, (4) an automatic recording device of propeller blade turning angles, (5) an automatic recording device of atmospheric pressure on each portion, (6) an airtight room (to install measurement devices), (7) a device defining attack angles, and (8) a defining device of load multiple from –7° to +15°.[57]

Figure 4.2 A flow line of the machine works of the 4th Special Design Bureau, OKB-4.
Source: RGAE F. 8044 Op.1 No. 6676. 52.

Figure 4.3 Piloting device test-bed of the laboratory in the 4th Special Design Bureau, OKB-4.

Source: RGAE F. 8044 Op.1 No. 6676. 54.

On July 3, the development research work at OKB-4 was classified into four task groups and scheduled to be conducted sequentially starting from the first task group. The first task group was intended for the development of controlling devices and probe-type compasses, which had already finished at this point. The second task group was intended for the design and manufacture of a piloting device driven by a motor (Figure 4.3) and a controlling gyroscope used in the gyroscope station (Figure 4.4), which was expected to be finished by December 31, 1946. The third task group was intended for the design and production of the two remaining gyroscopes that constitute the gyroscope station, the installation of already tested devices such as the absolute altimeter on the airplane, and the development of an autopilot device (Figure 4.5) as an assembled system. These tasks were expected to be completed by May 1, 1947 (not including the test flight schedule). The final task group was intended for the development of an automatic take-off and landing device, which was expected to be the most difficult, and scheduled to be completed by April 1, 1948. With regard to the transfer of OKB-4 to Soviet Union territory, Director Leont'ev advised that it be postponed till after May 1, 1947, when the third task group was expected to finish,[58] because "an early transfer would result in unexpected interruptions."[59]

Figure 4.4 Adjustment of gyroscope device by a qualified engineer at 4th Special Design Bureau, OKB-4.

Source: RGAE F. 8044 Op.1 No. 6676. 54.

Figure 4.5 Installation of a test-bed for inspection of autopilot system at 4th Special Design Bureau, OKB-4.

Source: RGAE F. 8044 Op.1 No. 6676. 56.

3 Transfer of the research and development centers

3.1 Transfer of machinery and German "Collaborators"

The Potsdam Agreement of August 2, 1945, prohibited the manufacture of ammunition within German territory.[60] Therefore, before entering the actual manufacturing stage, it was necessary to transfer the research and development centers out of German territory. It seems that the ministerial ordinance No. 228 of April 19, 1946, ordered to transfer the four OKBs to Soviet territory. In August, the Deputy Minister issued the directive No. 02 / 195s., which ordered the preparation of a list of equipment to be removed and transported (volume, maximum dimensions and weight, volume of packing material required for the transportation, and the number of freight cars or trucks) in addition to a report on the advancement of preparation. According to that directive, the following items were to be sent from OKB-1 to the first of the two experimental factories that would be established as new research and development centers within the Soviet territory: (1) all the press machines from machine works; (2) large polishing machines, fraises, and grinding machines; (3) all the equipment from the unit part works and airplane assembly lines; (4) the machinery and installation from coating material works; (5) non-ferrous metal casting equipment; and (6) a forging press machine. The items to be sent from OKB-3 were: (7) research equipment and instruments for measurements and control; (8) an airframe test station for a static experiment equipped with instruments for measurements and control; (9) an engine test-bed equipped with instruments for measurements and control; (10) a stamping and press machine for airplane parts; (11) specialized equipment; (12) an assembly station for airplane and unit parts; (13) the photograph cabinet of OKB-3; (14) reserve fuel storage; (15) stock of products, semi-finished products, raw material, fitting tools, and instruments for the manufacture of airplanes; and (16) winches and transferring devices. Furthermore, the following items were to be sent from OKB-1 and OKB-2 to the second experimental factory: (1) all high-level research equipment and all instruments and devices for measurement and control; (2) a hangar equipped with measuring instruments for engine tests; (3) a test-bed for jet engine tests equipped with measuring instruments; (4) material for jet engines, research equipment for the tests of unit parts; (5) large lathes, boring machines, and frontal lathes; (6) large fraises, drills, grinding machines, and polishing machines; (7) forging press machines; (8) devices for the tests of variable-pitch propeller roller bearing box and a test-bed equipped with a single cylinder fitting device; and (9) winches and transferring devices of OKB-1.[61]

A large-scale transferring operation covered not only research equipment such as the ones mentioned above but also manufacturing equipment considered necessary for full-fledged industrialization. The number of machines from German companies in Soviet-Union-occupied areas to be transferred to Soviet territory was enormous; the Ministry of Aviation Industry alone had jurisdiction over a total of 65,96 items: (1) 30,565 metal-cutting machine tools, (2) 4,921 press machines, (3) 79 hammers, (4) 5,467 electric motors, and (5) 27,928 units of various other equipment.[62] After making the decision to transfer machinery to the Soviet territory, on May 10, the representative of the Ministry of Aviation Industry in Germany requested reports detailing the number of equipment to be carried out from each German company. According to the summary of the reports dated June 1, there were 30 factories which were the targets of the transferring operation under the jurisdiction of the Ministry of Aviation Industry.[63] Table 4.1 lists the names of those factories along with the number of equipment that were to be carried out (as long as it contains specific information.)

Along with the equipment, the necessary number of German "collaborators" was also transferred. This number, as calculated when the research and development base was still in Germany, was approximately 8,000 people. However, Resolution No. 874–366ss issued by the Council of Ministers (on April 17, 1946) determined that only 1,400 or 1,500 selected persons would be transferred. This selection was due to finish in October,[64] but the list of 1,400 German specialists and workers (3,500 including their families) who would be transferred was already determined on August 7.[65] On October 2, 1946, more than 1,000 specialists were crammed into trains and transported to the Soviet territory overnight. The employees of Junkers in Dessau were transported to Upravrencheski in the north of Kybyshev (current Samara), and the members of Köthen Factory were transported to the design bureau of Klimov in Chernikovsk.[66]

3.2 Continuing research activities in Germany and "the Case of Dr. Keune"

However, it appears that many of the German specialists sent all the way to the Soviet Union did not have much work to do. This was because the preparation and provision of the research equipment to be used in the Ministry were so far insufficient for the needs issued by the Ministerial Purchase Commission. The basic reason behind this was the lack of qualified specialists and skilled workers and a disorganized industrial basis. Therefore, Deputy Minister Makar Lukin (1905–1961), stationed in Germany, advised Minister Khrunichev on the reorganization and continuation of the "experimental studies in the existing manufacturing bases (Dessau, Halle, Berlin, Stassfurt)." He added, "It may be useful to take advantage of the existing equipment so as to equip these manufacturing bases with research and other facilities, not bringing any important change to the equipment of the First and Second Experimental Works."[67] He also suggests concretely the facilities in every OKB that could be diverted. This advice, however, was not followed. Between November and December 1946, all the four OKBs were completely transferred to Soviet Union territory.[68]

Meanwhile, the engineering division in Germany of the Ministry of Aviation Industry, responsible for solving minor technical problems, was reorganized as the "Scientific and Technological Section (*Nauchno-Tekhnicheskii Otdel*; NTO)" of the Ministry of Aviation Industry in late 1946. It became affiliated simultaneously to the Ministry of Aviation Industry, the Directorate on the Study of German Scientific and Technological Achievements (*Upravlenie po izucheniyu*

Table 4.1 List of companies subject to the equipment transfer strategy under the jurisdiction of the Ministry of Aviation Industry

(1) Factory of "Argus" Company: 1,497 metal-cutting machine tools, 20 press machines, two hammers, 2,080 various other equipment, and 1,022 cases of instruments and tools were already loaded onto nine railroad freight wagons.
(2) Factory of "Hirth" Company (Waltersdorf, Bezirk Neukölln): 235 units of equipment.
(3) Factory of "Kurt Hedelman*" Company: 236 units of equipment.
(4) Factory of "Tevez* Werke" (Wittenau) Company: 783 units of equipment out of which 234 were metal-cutting machine tools.
(5) "Defauel*" Institute: 6,683 machines out of which 436 were metal-cutting machine tools.
(6) BMW Company, Spandau Factory: to be reported later.
(7) Henkel Company, Rostock Factory: 2,337 units of equipment (out of which, 938 were metal-cutting machine tools) were transported on 17 freight trains and one truck.
(8) Henkel Company, Oranienburg Factory.
(9) "Arado" Company, Rathenow Factory: all 1,024 units of equipment (out of which, 424 were metal-cutting machine tools) had already been dispatched to the Soviet Union. These included sanitary facilities and measuring and controlling instruments.
(10) "Arado" Company, Hönofen* Factory.
(11) "Arado" Company, Potsdam Factory.
(12) "Schlegel" Company, Golson* Factory.
(13) BMW Company, Basdorf Factory: 1,229 units of equipment.
(14) "MEMEFA" Company, Waren Factory: As 114 units were distributed to the People's Commissariat (then) of Non-Ferrous Metal Industry and 69 to People's Commissariat of Heavy Construction Industry (then), only 784 out of the 968 equipment units were transferred to the Ministry of Aviation Industry by 20 transportation corps.
(15) Factory of "Askania Werke": 1,240 units of equipment transferred to Factory No. 122.
(16) BMW Company, Halberstadt Factory: 755 units of equipment.
(17) "Eiso*" Company, Eisfeld Factory: 242 units of equipment.**
(18) "Gustloff Werke," Kahla Factory: 682 units of equipment.
(19) "Siebel Flugzeug Werke," Halle Factory: as experimental studies were still ongoing, the number of equipment to be sent was yet to be determined.
(20) "Junkers" Company, Dessau Factory: same as above.
(21) BMW Company, Stassfurt Factory: same as above.
(22) "Junkers" Company, Schönebeck Factory.
(23) "Rautal Werke," Wernigerode Factory.
(24) "Mitteldeutch Werke," Härtzgerode* Light Alloy Factory.
(25) "Elster" Company, Berstendorff* Factory.
(26) "Daimler-Benz AG," Berstendorff* Factory.
(27) "Junkers," Ebersbach Factory.
(28) "Venus Werke," Venusberg Factory.
(29) "Junkers," Mecklenburg Factory.
(30) Oberlungwitz Power Plant: a turbine, turbine generator, electric motor, pump, condenser, transformer, and vapor pump were transferred to Factory No. 39 in Irkutsk.

* Original spelling – unknown. Directly Romanized from the Russian pronunciation.

** It says "out of which 313 were metal-cutting machine tools," but this is likely to be an error. It is not known whether this should be the external number or if the number itself is wrong (Ichikawa).

Source: RGAE F. 8044 Op.1 No. 6671. 55–57.

dostizhenii nauki i tekhniki Germanii) of the government and the State Planning Committee (Gosplan). The NTO began its activities in 1947. As it did not have its own OKB or scientific technology bureau, the NTO tried to carry out the research activities by sealing contracts with and entrusting the research to Soviet Joint-stock Company [*Sovetskoe Aktsionernoe Obshchestvo* (German companies whose stocks the Soviet Union had acquired)], other German companies and individual German specialists. It was a measure aimed at managing efficiently the subjects remaining even after the transfer of the R&D bases, by reducing costs, improving the sense of responsibility of German engineers and workers, and releasing the few Soviet specialists from non-essential tasks related to the management of OKB. Although there used to be 24 subjects in the first year (1947), immediately after the start, two of them were discarded owing to the absence of specialists in Germany. Instead, many new subjects were added throughout the year. Moreover, the subjects were often expanded or combined with others. Eventually, their number settled at 42 by the end of 1947. Out of the newly added subjects, 14 were subjects related to hydrodynamics. Among those, the task of designing a wind tunnel for aerodynamic experiments at supersonic speed and automatic devices for aerodynamic weight scale was important, but very complicated. Moreover, there were 11 subjects related to aerodynamics theory, and most of them were intricate ones related to supersonic speed. For 30 of the subjects, 55 contracts were signed with the Soviet Joint-stock Companies (21 cases), German companies (24 cases), and individual specialists (10 cases). At the beginning, six Soviet specialists were allocated to NTO and its local organizations. However, by late 1947, their number increased to 11. Further, more German specialists were also mobilized as extra staff members; 343 German specialists engaging in 42 subjects totally were enlisted. However, the solution of 13 complicated subjects among the subjects enlisted as auxiliary ones required a stable research base, which led to the establishment – although irregular – of the Scientific and Technological Bureau, "Pribor" (*Nauchno-Tekhnicheskoe Byuro "Pribor"*; NTB-Pribor. Pribor means instruments, usually measuring and controlling instruments) within the NTO. The rate of achieved subjects among all the subjects in 1947 was 67.6%, which was slightly higher than the planned 67.3%, and a total of 454 types of models were prepared.[69]

Such an achievement, however, was far from satisfactory to P. Shnitman, Director of NTO. He mentioned totally nine "difficulties and defects": (1) confusion caused by inadequate preparations and thoughtlessness in the initial plan; (2) stagnant initial activities due to a delay (until late May) in the decision of technological requests (specifications, standards and others) by the Ministry; (3) some of the NTO workers did not satisfy the level of skill required for the tasks in terms of expertise and qualification (for example, there were no specialists in radar and wireless communication and in NTB-Pribor, there were no Soviet specialists in aerodynamics); (4) the critical shortage of Soviet specialists; (5) NTO and NTB-Pribor did not have Soviet civilian interpreters or typists and, therefore, the secrets could not be retained; (6) the shortage of German aerodynamics specialists in NTB-Pribor (this will be mentioned later);

(7) a fixed annual expenditure budget for NTO was not secured; (8) inadequate leadership in Soviet Joint-stock Company (some executives of the companies, Siemens-Schuckert, Siemens-Prania, and Isolator did not take responsibility over the appointed date of delivery and the supply of finished products); and (9) a complete shortage of distribution.[70]

In addition to these "difficulties and defects," an incident that made difficult the continuation of research activities in Germany broke out. In late 1947, one of the few specialists in aerodynamics and one of the leading figures in NTB-Pribor, Dr. Keune (original spelling and details – unknown, probably F. Keune), disappeared. The research in NTB-Pribor was focused on designing wind tunnels for aerodynamic experiments at supersonic speed, designing an automatic aerodynamics weight scale and a study related to aerodynamics theory. Until a specialist suitable for these subjects was sent from the Soviet Union, V.A. Shmelev assumed temporarily the role of Director of NTB-Pribor and Dr. Keune played the role as the leader on the German side. The repeated requests to send Soviet specialists were not honored, and there was a serious shortage of German specialists in the Soviet occupied areas. Dr. Keune promised to persuade a few specialists living in the UK-occupied city of Göttingen to work in NTB-Pribor. Accordingly, having received a pass, he left toward Göttingen on November 28. Despite his promise to return between December 2 and 4, he did not. His family had already escaped by November 4. Shnitman, Director of NTO, recommended the Chief on Human Resources of the Soviet Military Administration in Germany that, "It has become impossible to maintain the activities at NTB secret. Therefore, it is inappropriate to allow NTB to continue its activities within German territory."[71]

His recommendation was followed. In January of the following year, the activities of NTO came to a turbulent end. In a letter dated January 24, addressed to the Deputy Minister of Ministry of Aviation Industry, the plenipotentiary representative of the same ministry in Germany, Ozimkov (first name and patronymic unknown) and Shnitman reported that NTO had completed the preparation of the models of machinery and measuring and controlling instruments and that the transfer to the Soviet territory was being prepared.[72] Subsequently, in a letter from Ozimkov[73] dated March 19, there was a retrospective note saying that NTB-Pribor had already shut down, which suggests that the R&D activities in Germany ended in early 1948.

4 The first military jet planes of the Soviet Union

4.1 German-made engines and their improvement

From late 1944 to early 1945, all the famous aircraft designers, such as Artyom Mikoyan, Aleksandr Yakovlev (1906–1989), Lavochkin, Sukhoi, Sergei Ilyushin (1894–1977), and Andrei Tuporev (1888–1972) started developing jet planes. As mentioned before, the domestic engines developed by Lyul'ka and his team were not so reliable. Therefore, they opted to use the engines obtained from Germany as "war trophies." The Jumo-004 with traction of 850 kg was newly codenamed

RD-10 and the BMW-003 with traction of 800 kg as RD-20.[74] The experimental studies on Jumo-004 and BMW-003 were conducted at the First Scientific Research Institute (*Nauchno-issledovatel'skii institute*; NII-1) and the Central Institute of Aviation Motors (*Tsentral'nyi Aviatsionnyi mashinostaritel'nyi institute*; TsIAM) of the Ministry of Aviation Industry and subsequently transferred to serial production in Factories Nos. 26 (Ufa) and 16 (Kazan').[75] The design bureau of Chetverikov (first name and patronymic unknown) engaged in the design of a bomber equipped with four BMW-003 engines and another bomber with two Jumo-004 engines on the basis of Arado-234. In factories numbered 82 and 458, Arado-234 and Messerschmitt:(Me)-262 were reproduced. In 1946, 120 units of Me-262 were manufactured in Factories Nos. 381 (Moscow) and 292 (Saratov).[76]

However, according to the recollection of Yakovlev, who was an aircraft designer and, then, Deputy Minister of Aviation Industry, their initial target was to copy the Me-262, but the complexity in control and a setback in the transfer to serial production led to the failure of this path. This is definitely a reflection of the incomplete technology acquisition we already analyzed in the previous part of this chapter. Hence, suddenly, the path of developing "domestic" planes, utilizing the technological information of German-made Jumo-004 and BMW-003 engines, was chosen.[77]

The MiG-9, which had a test flight on April 24, 1946, and became the first jet plane of the Soviet Union, was equipped with two units of RD-10. The Yak-15, which had a test flight shortly afterward, was equipped with a RD-10, i.e., a Jumo-004.[78] The MiG was equipped with a BMW engine and Yak with a Jumo engine. The BMW engine was smaller than the previously mentioned engine developed by Lyul'ka for the same plane and hence, the MiGG team had to reduce the size of the airframe.[79] The jet plane was first revealed in the May Day parade of 1947.[80] Subsequently, La-150, Su-5, and Su-9 were finished on the basis of RD-10 engine[81] (Figure 4.6).

Furthermore, the turbo prop engines, Jumo-012 type were manufactured by the German engineers Brandner and Scheibe in Kybyshev. Jumo-012 was renamed to NK-4 and subsequently equipped to Oleg Antonov (1906–1984, an aircraft named after its design bureau, abbreviated as An-8 [an aircraft named after an eminent designer, Oleg Antonov (1906–1984)] as NK-6, immediately after some improvement by Nikolai Kuznetsov.[82]

The German engine Walter was useful in the improvement of RD-1.[83] An Il-22 bomber was equipped with four TR-1 engines, which was the first Soviet-made engine designed by Lyul'ka. Its first test flight was carried out on July 24, 1947.[84] Su-11 was also equipped with a TR-1 (traction of 1,250 kg), but this type was not transferred to serial production. Subsequently, under Lyul'ka's leadership, the engines TR-2, TR-3 (Akh-5, traction of 5,100 kg), Al-7, and Al-7f (9,600 kg) were transferred to the serial production mainly for Sukhoi planes. Following his footstep, Klimov engaged in the development of domestic engines and Aleksandr Mikulin (1895–1985) developed an AM-3 with traction of 8,750 kg in 1951 and a RD-3m (1,120 kg) in 1952. Sergei Tumanskii (1901–1973) developed an RD-9b (3300 kg) for fighter aircrafts in 1953.[85]

Figure 4.6 Line-up of the Soviet main warplanes immediately after the war.

1. A fighter, Yak-9; 2. A fighter, Mig-15; 3. A front Bomber, Il-28; 4. A long-range bomber; 5. A supersonic multi-purpose airplane, Yak-28; 6. A long-range bomber, Tu-16; 7. A supersonic fighter, Mig-21; 8. A supersonic fighter=bomber; Su-7b.

Source: *Sovetskaya voennaya entsiklopediya*. Tom 1. M.; Voenizdat, 1976. 53.

It appears that the leadership of the Party and the government expected quick results. As early as mid-1946, Stalin's anger about the "delay" erupted in the form of criticism against the bureaucratism and sectionalism of Director of the Main Directorate of Artillery, Nikolai Yakovlev (1898–1972, not the same person as the aircraft designer Yakovlev), Minister of Aviation Industry, Aleksei Shakhurin, and the General Commander of the Air Force, Aleksandr Novikov.[86] As mentioned before, the Soviet engineers labored immensely with the development of the "domestic planes," especially to guarantee heat resistance. Even Mstislav Keldysh, who would later become President of the USSR Academy of Sciences, could only propose the solution of separating the engine from the airframe slightly.[87] Moreover, as the speed increases, determining the best-suited wing shape becomes difficult. As for the design research for determining it, Georgii Svishchyov (1912–1999) and Ya. M. Serebritskii proposed a new wing model and Vladimir Struminskii (1914–1998) conducted a theoretical study on it. Such efforts were crystallized in MiG-15 (traction of 4,800 kg, maximum speed 1,050 km/h, equipped with one 37mm and two 23mm cannons), which played a very important role in the Korean War. The La-174, tested in January 1948, broke the barrier of sound for the first time. Moreover, in 1950, the MiG-17 attained the speed of Mach 1.05. In 1952, the MiG-19, equipped with two RD-9b engines (traction of 3,250 kg) by Tumanskii, three 30mm cannons and with airframe weight of 7.5 tons, achieved a maximum speed of 1,450 km/h and was serialized. In the same year, a Yak-25 equipped with two units of RD-9 (traction of 2,600 kg) flew for 3,000 km at 1,090 km/h. It had an airframe weight of 9.2 t and two 37 mm cannons. The most important feature, however, is that this aircraft had a built-in radar.[88]

4.2 Introduction of British technology, Tuporev and Ilyushin

Along with the "plunder" of German technology, an extremely well-timed licensed production played a significant role in the transfer of jet engine technology to the Soviet Union.

In September 1946, the British Labour administration lifted the ban on technology exportation to the Soviet Union and authorized Rolls-Royce to export jet engines to the Soviet Union. The shipping of 30 Derwent engines and 25 Nene engines started in March 1947 and continued through the following year. Moreover, after observing an Air Show in Paris immediately after the war, Klimov traveled to London to negotiate about the license to produce British jet engines and, consequently, he obtained a license from Rolls-Royce in 1947.[89] Thus, the Derwent engine with traction of 1,600 kg began to be used under the Russian codename of RD-500 and the Nene engine of 2,200 kg as RD-45.[90]

In late 1946, according to a proclamation of the Council of Ministers, Tuporev started a development research on a bomber equipped with jet engines and completed its drawing in the summer of 1947. This became the Tu-12.[91] It is said that Tuporev was expecting the British engines for this stage.[92] The Tu-12, which was the first bomber equipped with jet engines manufactured in the Soviet Union, was equipped with an RD-45 (although it was not mass-produced).[93] It had a maximum speed of 783 km/h and altitude of 11 km. Further, the government and the Central Committee of the Communist Party ordered Ilyushin and Tuporev to develop a bomber for frontline equipped with a jet engine. This order was realized as Il-28 and Tu-14. Tu-14 was known for its unique design with a third engine attached to the rear of its airframe and exhibited good performance in a test in late 1947.[94] Tu-14, which had its first flight in 1948 and was subsequently transferred into serial production, was initially mounted with an RD-45 engine, but later with a VK-1 – a version improved by Klimov. RD-500 was manufactured in Factory No. 500 in Tushno (now a part of Moscow) and the RD-45 in Factory No. 45 in Moscow. The latter was immediately improved by Klimov and loaded to MiG-15 under the name VK-1. Eventually, as many as 5,000 units of this aircraft would be produced.[95]

From 1946 to 1947, Ilyushin designed the Il-28. It was initially equipped with two RD-45s, which were subsequently replaced by VK-1. It had a flight range of 2,400 km at 900 km/h with the traction of 2,000 kg. Its airframe weighed 21.2 tons. Il-28, which could be loaded with 1,000 kg of explosive, was equipped with four 25mm cannons and could carry three people, was immediately selected as the standard frontline bomber. From 1948 to 1949, the design of Tu-16 started. According to Mikulin, equipped with three AM-3 engines with traction of 8,750 kg, it had a maximum speed of 1,000 km/h and flight range of 3,700–5,700 km, carried a 23-mm cannon and had the capacity to carry six people.[96] As soon as its test flight was conducted successfully in 1952, it was transferred to serial production. One of the problems was that its maneuverability decreased at the high-speed of approximately 900 km/h. A group led by Sergei Khristianovich (1908–2000) at TsAGI and another group led Ivan Ostoslavskii (1904–1972) and M.A. Gaits (first name and details unknown) at the Flight Research Institute (Lyotno-issledovatel'skii institute; LII) investigated it in parallel. Moreover, a German group led by Baade completed the design of the bomber B-140 in 1949. The maximum speed attained was approximately 1,230 km/h.[97]

Thus, by the completion of the previously mentioned Yak-25, the first "main four models" of the Soviet Air Force (main fighter aircraft MiG-19, main interceptor Yak-25, main close-range bomber Il-28, and main long-range bomber Tu-16) were finally completed.[98]

Notes

1 L.L. Kerber, *Tuporev*. St. Petersburg, Politekhnika, 1999g, 248.
2 I.G. Sultanov, *Istoriya Sozdaniya Pervykh Otechestvennykh Tuboreaktivnykh Samolyotov*. Moscow, Vuzovskaya kniga, 1998, 6, 7.
3 N.S. Stroev, "Ch. 6. Voennaya aviatsiya." in A.V. Minaev (Ed.), *Sovetskaya Voennaya Mosh' ot Stalina do Gorbacheva*. Moscow, Voennyi parad, 1999, 271.
4 Yak-15 finally achieved 805 km/h (on transferring to serial production, it decreased even further to 786 km/h) with the traction of 900 kg, which shows that this estimation was overoptimistic. In fact, LaGG-3 succeeded for the first time in 1946, when RD-1 was improved based on German jet engines (Sultanov, *op. cit.*, in note 2. 10).
5 Surtanov, *op. cit.*, in note 2. 8–10.
6 Kerber, *op. cit.*, in note 1. 248.
7 U. Albrecht, *The Soviet Armament Industry*. Chur, Switzerland, Harwood Academic Publishers, 1993 (Originally published in German in 1989 as *Die sowjetische Rustungsindustrie* by Westduetscher Verlag, Opladen, Germany), 43.
8 D.A. Sobolev, *Istoriya Samolyotov 1919–1945*. Moscow, ROSSPEN, 1997g, 314.
9 Ibid.
10 Ibid.
11 Ibid., 316.
12 Ibid., 317, 318.
13 A.S. Yakovlev, *Tsel' Zhizni: Zapiski Aviakonstruktora*. Moscow, Politicheskaya Literatura, 1968.
14 Albrecht, *op. cit.*, in note 7. 35.
15 Rossiiskii gosudarstvennyi arkhiv ekonomiki (Russian State Archive of Economy. Hereafter, RGAE). Fond (F.) 8044, Opis' (Op.) 1, No. 6683, 13, 14.
16 Some documents (for example, the references number 23 and 45 of this chapter) often refer to this OKB as "Osoboe Tekhnichekoe Byuro (Special Technology Bureau; OTB)," but here we unified it as OKB, the most common nomenclature.
17 RGAE, F. 8044, Op. 1, No. 6679, 17, 18.
18 Albrecht, *op. cit.*, in note 7. 30–33, 36.
19 RGAE, F. 8044, Op. 1, No. 6679, 7.
20 RGAE, F. 8044, Op. 1, No. 6683, 14, 15.
21 Ibid., 17.
22 Ibid., 25, 26.

23 RGAE, F. 8044, Op. 1, No. 6679, 9.
24 RGAE, F. 8044, Op. 1, No. 6683, 20, 21.
25 RGAE, F. 8044, Op. 1, No. 6679, 10.
26 PRGAE, F. 8044, Op. 1, No. 6683, 21.
27 Ibid., 22.
28 Ibid., 24, 25.
29 RGAE, F. 8044, Op. 1, No. 6675, 6.
30 Ibid., 28.
31 RGAE, F. 8044, Op. 1, No. 6678, 23.
32 Ibid., 29.
33 RGAE, F. 8044, Op. 1, No. 6678, 81.
34 Ibid., 23, 24.
35 Ibid., 82.
36 RGAE, F. 8044, Op. 1, No. 6679, 18.
37 RGAE, F. 8044, Op. 1, No. 6678, 24.
38 Ibid., 29.
39 Ibid., 83.
40 RGAE, F. 8044, Op. 1, No. 6683, 19.
41 RGAE, F. 8044, Op. 1, No. 6677, 3.
42 Ibid., 73.
43 Ibid., 10.
44 Ibid., 6.
45 Ibid., 4.
46 RGAE, F. 8044, Op. 1, No. 6683, 23.
47 RGAE, F. 8044, Op. 1, No. 6677, 7.
48 Ibid., 5.
49 Ibid., 73.
50 RGAE, F. 8044, Op. 1, No. 6683, 23.
51 RGAE, F. 8044, Op. 1, No. 6677, 7.
52 Ibid., 75.
53 RGAE, F. 8044, Op. 1, No. 6676, 121, 123.
54 Ibid., 45.
55 Ibid., 121, 122: in the original language, "gyroscope" was often referred to as "tri-(ring) gyroscope (tryokhzhiroskop)," but as a regular gyroscope has a structure with a rotor inside three metal rings rotating around mutually vertical axes, here the "tri-(ring)" was abbreviated. Further, there are no documents that describe the "probe-type compass (zondnyi kompass)," which is made of a "Mutter (mother)-compass" and a "Tochter (daughter) compass."
56 RGAE, F. 8044, Op. 1, No. 6683, 30–35.
57 Ibid., 36, 37.
58 RGAE, F. 8044, Op. 1, No. 6676, 49.
59 Ibid., 48, 49.
60 Albrecht, *op. cit.*, 33.
61 RGAE, F. 8044, Op. 1, No. 6667, 45.
62 RGAE, F. 8044, Op. 1, No. 6671, 116.
63 Ibid., 55–57.
64 RGAE, F. 8044, Op. 1, No. 6689, 6.
65 Ibid., 12.
66 Albrecht, *op. cit.*, in note 7. 33.
67 RGAE, F. 8044, Op. 1, No. 6689, 6–8.
68 RGAE, F. 8044, Op. 1, No. 6690, 2.
69 Ibid., 2–6.
70 Ibid., 11–13.

71 RGAE, F. 8044, Op. 1, No. 6695, 45–47.
72 PRGAE, F. 8044, Op. 1, No. 6701, 12.
73 Ibid., 54: This letter was meant to ask whether TsAGI required the technical documents left behind after the shutdown of NTB-Pribor or not. Here, the title of Ozimkov was written as "Director of NTO."
74 Stroev, *op. cit.*, in note 3. 271.
75 Albrecht says that the production of BMW-003 started in 1947 in the Factory No. 466 "Red October" in Leningrad and that Jumo-004 was produced in the Factory No. 10 in Kazan' (Albrecht, *op. cit.*, in note 7. 36), but this probably should be corrected.
76 N.S. Simonov, *Voenno-promyshlennyi Kompleks SSSR v 1920–1950-e gody: Tempy Ekonomicheskogo Rosta, Struktura, Organizatsiya Proizvodstva i Upravlenie*. Moscow, ROSSPEN, 1999g, 236.
77 Sultanov, *op. cit.*, in note 2. 10–11.
78 Stroev, *op. cit.*, in note 3. 271.
79 Albrecht, *op. cit.*, in note 7. 44.
80 Sultanov, *op. cit.*, in note 2. 5.
81 Stroev, *op. cit.*, in note 3. 272.
82 Albrecht, *op. cit.*, in note 7. 35, 36.
83 Simonov, *op. cit.*, in note 76. 236.
84 Sobolev, *op. cit.*, in note 8. 58.
85 Stroev, *op. cit.*, 271–272.
86 Simonov, *op. cit.*, in note 76. 238: Shakhurin was dismissed as Minister on January 5, expelled from the Central Committee of the Communist Party in March and finally arrested on April 4. The Supreme Court sentenced him to seven years in prison for "abuse and excess of authority in an aggravating situation" and "manufacturing irregular of products of poor quality that do not satisfy the required standards" (K.A. Zalesskii, *Imperiya Stalina: Biograficheskii Entsiklopedicheskii Slovar'*. Moscow, VEChE, 2000g, 488). Novikov was also imprisoned in almost the same period (ibid., 338). Nikolai Yakovlev was spared from prison, but was discharged in 1951 (ibid., 514).
87 Sobolev, *op. cit.*, in note 8. 28, 29.
88 Stroev, *op. cit.*, in note 3. 272–279.
89 Albrecht, *op. cit.*, in note 7. 19, 26, 37.
90 Stroev, *op. cit.*, in note 3. 271.
91 Kerber, *op. cit.*, in note 1. 250.
92 Albrecht, *op. cit.*, in note 7. 26.
93 Stroev, *op. cit.*, in note 3. 271.
94 Kerber, *op. cit.*, in note 1. 250, 253.
95 Albrecht, *op. cit.*, in note 7. 38.
96 Stroev, *op. cit.*, in note 3. 272.
97 Sobolev, *op. cit.*, in note 8. 60.
98 Stroev, *op. cit.*, in note 3. 279.

5 “Concealed rivalry”

The early days of computer development

Strela or BESM: Which computer was developed first in the Soviet Union? Which one was first successful? Such questions are difficult to answer, but recently declassified archival documents may hold the key. The story of the Strela illustrates how competing interests – as institutions and factions jockeyed to gain political advantage – helped determine the fate of this computer, a political success but a technological failure.

1 Early Soviet development focus

What were the main subjects of the computing device developments at that time? Nikolai Bruevich (1896–1987), Natan Kobrinskii (1910–1985), Lev Gutenmakher (1908–1981), N. A. Borodachev, and Lazar’ Lyusternik (1899–1981), who were later to be leading staff of the Institute of Precision Mechanics and Computing Technique (*Institut Tochnoi Mekahniki i Vychislichel’noi Tekhniki.* Hereafter, ITMiVT), wrote, in June–July 1946[1]:

> That which occupies a special position in computing techniques is the solution of a variety of mathematical subjects related to overland or oversea aeronautics, firing in artillery, precision bombing meters, anti-aircraft guns and firing of long-distance missiles. They will be realized with special devices (PUAO, PUAZO and others) and other aiming devices. . . . The research will be carried out with the purpose of the development of new calculating and solving machines, in which the integrators for the general differentiate equations, the machines for the algebraic equations, the electrical integrators for the frontiers in the partial derivatives with the application of the differentiate equations, the machines for the high-order algebraic equations, and the multi-purpose punch-card calculating machines, are to be included.

ITMiVT was set up in July 1948, by an order of the Presidium of the Academy of Sciences, with approximately 60 researchers, most of whom came from the Institute of Mechanical Engineering, Energetics Institute, and V. A. Steklov Institute of Mathematics.[2] Ministry of Machine-building and Instrument-building (*Ministerstvo Mashinostroeniya i Priborostroeniya.* Hereafter, MMiP) offered the site of 830 m^2 of one of its factories to that newly founded institute.[3] The President of the Academy of Sciences, Sergei Vavilov (1891–1951), recommended Nikolai Bruevich, a specialist of bombing-meters and lieutenant general, for the position of the first Director of the Institute.[4]

Reporting the activities of ITMiVT in its first year (1948), Bruevich wrote,[5]

> We embarked on the research for the development of general methods for tabulating of the multi-variable functions, which will be adapted for the subjects of ballistics, geodesy, radio-navigation and others. . . . Moreover, a sample of Eli-11, the integrator equipped with electronic tubes for a system of continuous differentiate equations, was developed and built up. . . . Some of our achievements were adapted at the Ministry of Aviation Industry and other specialized research institutes.

An engineer who was present at that time says that Bruevich, in his capacity as Acting Director, brought up the matter of developing a general-purpose digital computer in a seminar held in early 1949. His proposal, however, met with strong objections from his colleagues. According to the witness of P. P. Golovistikov, Bruevich's colleagues raised three objections:[6]

> First, the vacuum tubes were so unreliable and the quantity of tubes required for any potential machine was so great that the machine might not be so reliable, while analog machines had already proven quite reliable. Second, even if it were shown that universal electronic digital machines could function reliably, given the probable high cost of research and development, it would be more cost effective to continue to develop special-purpose machines. Finally, the expertise required to operate such machines seemed overwhelming.

Therefore, the Institute's Plan of Scientific Research Activities for 1949 did not make any mention of a general-purpose digital computer.[7]

On the other hand, Special Design Bureau No. 245 (*Spetsial'noe Konstruktorskoe Byuro No. 245*. Hereafter, SKB-245) was established by the Soviet Council of Ministers in December 1948, soon after the establishment of ITMiVT. Its duty was to develop electrical analogue calculators and precision instruments, and improve existing special-purpose calculating devices, especially needed for designing rockets and jet planes. The Bureau began its service with two engineers, but within just a year it recruited as many as 144 people, of whom 97 were engineers, technicians, or men with technical expertise.[8] The others served in the administrative sections. Nearing its peak, by 1951, the number rose to 378.[9] By then the Bureau was running 12 laboratories and an affiliated factory in Moscow.[10] Soon the Bureau also began to operate another factory, called the Penza "SAM" (*Schetno-Analiticheskaya Mashina*; Calculator-Analyzer). The Penza SAM itself helped establish a subsidiary of SKB-245, which employed 50 men, most of whom were specialists.[11]

But all did not go so smoothly for the Bureau. For the fiscal year of 1949, the estimated budget for the Bureau was about 7 million rubles, but the amount it finally expended was far smaller: 4,636,600 rubles. At that time the Bureau could not reach its full quota of work because of a shortage of specialists. Furthermore, the contractor-clients, all government institutions, were neither diligent nor well informed about computing devices and, therefore, they very often delayed sending their specifications to SKB-245.[12] Such delays meant difficulties in the Bureau's financing, because about half of the Bureau's finances came directly from the contractors. For example, in 1952, the share of the direct government contribution to the Bureau's budget was about 16,150,000 rubles of the total budget of 30,950,000 rubles. The rest came from government contracts.[13]

The high productivity of the Bureau at this time was largely due to the preference of its clients, in particular, of the sections of the military mainly related with trajectory and aeronautical dynamism, for special-purpose computing devices to universal computers.[14]

2 The objections from the "forward-looking" mathematicians

The Soviet Academy of Sciences led some of the crucial research in precision mechanics and computers in the Soviet Union.[15] However, according to the memoirs of a famous mathematician, Mikhail Lavrent'ev (1900–1980), who was later to be Vice President of the Academy of Sciences, most scientists and engineers at the Academy underestimated the potential of computers:[16]

> After the appearance of computers in America, there arose a split in the opinion among our mathematicians, electric engineers and machinists. The majority of them considered a computer as an advertisement, without any prospects, and instead proposed to design more analogue computing machines and mechanical calculators. For this end, a new research institution, the Institute of Precision Mechanics and Computing Technique, was established.

As it is mentioned above, the Institute's Plan of Scientific Research Activities for 1949 did not make any mention of a general-purpose digital computer.[17] This Plan disappointed some mathematicians. Soviet scientists such as Lavrent'ev and Mstislav Keldysh (1911–1978), a great mathematician who later became President of the Academy of Sciences, were aware of the need for faster machines than those available.

What motivated Lavrent'ev and his colleagues?

Lavrent'ev, Keldysh, and other mathematicians were doing a tremendous number of complicated calculations needed for nuclear weapons development. In the Institute of Mathematics, where Keldysh worked, some "brigades" of calculators were engaged in such calculations. The brigades worked on a shift for eight hours for a working day. Each of them consisted of more than 100 people, most of whom were young girls. They used the German-made calculators, "Mercedes"-es.[18] Furthermore, in order to coordinate tremendous amounts of calculating tasks related to nuclear development, the First Main Directorate set up in its organization the Interdepartmental Commission on Computing Technique.[19] The level of the required calculations was so complicated, far beyond the level the modern artillery required in the trajectory research, that they were near the limit of their ability. A research associate of ITMiVT, Neiman (his given name and patronymic are unknown) reported at the Institute meeting of June 26–27 that, "the mathematicians are now presented by the physicists with the tasks which are very difficult to solve."[20]

In addition, according to Slava Gerovitch,[21] they were aware of the developments in computing in the United States and the United Kingdom. So, as early as 1947, Lavrent'ev referred to the importance of the development of computing mathematics at a discussion in the Department of Physical-Mathematical Sciences of the Academy of Sciences.[22]

The nationwide ideological campaign in the Soviet Union, however, made it hard for people to absorb foreign news, and adopt foreign ways and inventions.[23] To avoid harassment, scientists and engineers like Lavrent'ev and Keldysh had to be cautious with their professional pursuits. They had to prove the Soviet originality of their attempts. They were, however, already on the way to a settlement. A trial for the development of a Soviet-made computer was already set in Kiev where Lavrent'ev had long worked.[24]

Sergei Levedev (1902–1974), one of the Soviet electrical engineers, and his young colleagues devoted themselves to the development of computers independently from others in Kiev. Lavrent'ev, who also held the position of Vice President of the Ukrainian Academy of Sciences, supported them.[25]

As soon as the basic work for the first Soviet electronic-digital computer, named MESM (*Malaya Elektronnaya Schetnaya Mashina*, or a small electronic calculating machine) was completed, perhaps, at the end of 1949, Lavrent'ev showed it to a group of top political leaders in the Ukraine. They included Nikita Khrushchev (1894–1971), then Secretary of the Central Committee of the Ukrainian Communist Party, and Andlei Grechko (1903–1976), Commander-in Chief of the Ukrainian Army (later Minister of Defense for the Soviet Union).[26] Khrushchev was to leave Kiev for his new position in Moscow in December 1949.

Furthermore, Lavrent'ev came to feel the necessity to make "a certain highly authorized governmental organization" interested, and looked for a way to make a personal approach to Boris Vannikov (1897–1962), Director of the First Main Directorate of the Council of Ministers, who was in charge of developing nuclear weapons. In addition, he wrote letters to the Central Committee of the Party and the Council of Ministers, asking for favorable treatment for his Institute.[27]

Archival sources all keep silent about the details of the political maneuvers. However, it is guessed that in Moscow Khrushchev worked for Lavrent'ev's advantage, which in turn led to the downfall of Bruevich.[28]

3 Downfall of Bruevich: the turning point for the Academy of Sciences

Lavrent'ev and others began criticizing the Plan of ITMiVT for the year of 1949, which did not include the subject of digital computer development. Furthermore, they later succeeded in obtaining approval from many of the academicians enrolled in the Department of Technological Sciences of the Academy of Sciences for their proposal to inspect the activities of ITMiVT.

The Department set up a special commission, which consisted of such members as Aleksei Ilyushin (1911–1998), Ivan Oding (1896–1964), Dmitrii Panov (1904–1975), Bogdanov (his given name and patronymic are unknown), with Keldysh as the Chair. "The Conclusion"[29] of this Commission, which was adopted with the approval of all the members except for Oding, heavily criticized the disregard to the significance of developing general-purpose digital computers and the incoherent character of the Institute. "The Conclusion" includes "the Reference Opinion" of the Commission, blaming Bruevich for the lack of qualification in specialty.[30] Bruevich and others prepared their own "Reply"[31] to the Commission. In the meeting on July 6, 1949, of the Bureau of the Department of Technological Sciences of the Academy of Sciences, the "Conclusion" and the "Reply" were discussed.

As a conclusion of this session, the Bureau chose to compromise, saying that ITMiVT should take the obligation of developing general-purpose, high-speed computers as well as electrical, or mechanical, mathematical machines and other subjects; the activities concerning the application of general-purpose, automatic computers, the solution of the precision problems of mechanical machines and mechanisms, and the new methods for the solutions of mathematical subjects in the calculator-analyzers and the mathematical machines, etc.[32] The Bureau adopted the proposal made by A. B. Chernyshev to add to the Bureau's Proclamation the phrase: "Putting a special stress onto the development of digital, high-speed computers, [the Bureau] proposes that the Institute (ITMiVT) add the subject of the development of such machines to its plan and make progress in that subject, employing all possible means."[33] Bruevich, however, also took the responsibility for the development of the former, temporarily.[34]

Bruevich was not appointed full Director, so he remained as Acting Director for more than a year. Meanwhile, his Institute could not be fully established. In the fall of 1949, the Academy's recommendation of Bruevich to the position of Director of ITMiVT was withdrawn.[35] Lavrent'ev was, however, appointed Director in the place of Bruevich on March 15, 1950.[36] It seems that his above-mentioned political maneuver took more time. So, despite such a dishonor, Bruevich maintained his influence in the Institute for a while. Even after the appointment of Lavrent'ev as Director, Bruevich kept the post of Chief of the Department of Precision Mechanics of the Institute.[37]

On Lavrent'ev's request, Vitalii Ditkin (1910–1987) was appointed Acting Deputy Director in the place of Kobrinskii in the meeting of the Bureau on April 11.[38] Consequently Lavrent'ev proposed new members of the Scientific Council of the Institute, namely the appointments of Sergei Lebedev as the Chief of Laboratory No. 1, and Ditkin as Chief of the Experimental and Calculating Laboratory. Such a rapid reform by Lavrent'ev, however, caused some questioning.[39]

In the meeting of the Bureau of the Department of Technological Sciences on July 13, the Five Year Plan and the Plan of the Scientific Research Activities for 1951 of the ITMiVT, which were newly prepared under the leadership of Lavrent'ev, were examined. Although Bruevich, whose name was never referred to on any page thereof, remarked very angrily concerning the Plan, it was adopted.[40]

Consequently on July 26 and 27, 1950, a large-scale meeting was held inside the Institute,[41] in which the former activities of the Institute were severely criticized. Ditkin read the questions submitted to the leadership of the Institute one after another:

> What are the causes of the miscalculations in the solutions of the problems with the calculating and the analyzing machines? How long can the calculations go on continuously without mistakes? What kind of miscalculations are there? What is the percentage of working day when the machines are able to work without miscalculating? What is the ratio of the calculations not employing any mechanical means to those employing the help of cradle calculators and tabulators?

He revealed the low level of the computing technology of the Institute, answering the questions: "The rate of miscalculation is very high, at 2–5%; if the machine makes more miscalculations, we must give up using it. The ratio of the calculations without any mechanical means to those with mechanical helps is 1 : 8, or 1 : 10."[42] Moreover, he showed an example of a machine called D-11:

> Recently we obtained a calculator named D-11. That machine is a trophy. It lacks manuals. We cannot help learning it by practice. We don't know the mechanism of the machine. We don't know its maximum calculating ability, or how many relays can work at the same time.[43]

That was approximately one year after the meeting of the Bureau of the Department of Technological Sciences of the Academy of Sciences, in which the first step toward the development of high-speed computers had been made. The meeting adopted a resolution which aimed at the "building-up of strong, high-speed digital computers."[44]

In the fall of 1950, Bruevich, Kobrinskii, and their colleagues, who had been dispatched from the Institute of Mechanical Engineering, were all removed from their positions in ITMiVT on account of the fact that their qualifications did "not meet the needs of the required project and the political requirement," and they went back to the Institute, to which they formerly belonged.[45] At the same time, Dmitrii Panov, a professor at Moscow State University, was recommended as Deputy Director,[46] and he took up that position at the end of October.[47] Soon after that, the days of Bruevich, who had inspired Soviet scientists with awe of himself as an informant to political power for a long time,[48] came to an end: He was virtually banished from the Academy of Sciences.[49]

Immediately after Panov's appointment as Deputy Director of ITMiTV, he was asked for advice on the draft of the Proclamation of the Council of Ministers, which would show guidelines on the future development of computers. He wrote his opinion very quickly, taking only one hour and writing by hand without writing any draft. There he pointed out that the definition of "a high-speed numerical computing machine" in the draft of the Proclamation was very vague, and he said there were so many subjects that he was afraid that they were spreading their efforts too widely, and that the new subjects were too complicated for MMiP to take charge of, despite the fact that the MMiP had played a great role and the Academy of Sciences had also played an additional role.[50] His advice was consequentially taken. In November 1950, the Council of Ministers gave the Institute full responsibility for developing a universal automatic high-speed numerical computer.[51] The Proclamation of the Council of Ministers dated May 11, 1951, designated the date of the completion of that machine as April 1953.[52]

Levedev moved to Moscow, where he took charge of the development of a computer that came to be known as BESM, the first "successful" Soviet computer[53] (Figure 5.1). Additionally speaking for the fairness, even when the State Commission of 1955, headed by Keldysh, evaluated BESM very highly, the Commission reported that BESM was inferior to the American machines, for example IBM's NORK Machine. In addition, it had some serious defects. Nevertheless, it recommended its manufacture on a commercial basis, after some improvements.[54] SKB-245, meanwhile, gave the Commission a report that raised doubts about the originality of BESM. It claimed that BESM was "a mere copy

Figure 5.1 BESM-1, the pioneer of a series of Soviet numerical electronic computers, designed by the Institute of Precision Machinery and Computing Techniques.

Source: Akademiya nauk, Ordena Lenina Sibirskoe otdelenie, *Vek Lavrent'eva*, Novosibirsk, Izdatel'stvo SO RAN, 2000. 61.

of American machines or Soviet machines," and pointed to its shortcomings, for example, of not having germanium and ferrite, and therefore the problem of automatic programming was not yet solved.[55]

4 "Strela," or the Ministry's quest for monopoly and its failure

Despite the fact that the Institute of Precision Mechanics and Computing Technique was radically reformed, the Special Design Bureau No. 245 remained unchanged. The Bureau went on producing analogue or mechanical calculators for their clients.

A young colleague of the Bureau, Bashir Rameev (1918–1994) once wrote a draft for the design of a digital electronic computer as early as August 1948 in collaboration with Isaak Bruk (1902–1974)[56] who had been engaged in the design of some computers in the Energetics Institute and later designed small computers for limited usage, M-1, M-2, etc.[57] Nevertheless SKB-245 did not itself initiate a project for a high-speed computer until the above-mentioned reorientation in the Department of Technological Sciences of the Academy of Sciences. Then, first of all, the Bureau had to recall Rameev from his service in the army in the Far East.[58]

In those days Gutenmakher proposed to build a high-speed computer equipped with electromagnetic non-contact relays in the place of vacuum tubes. His proposal attracted Minister of MMiP Pyotr Parshin (1899–1970), perhaps because the Soviet Union was then suffering from a shortage of vacuum tubes.[59] His proposal was, however, denied in a meeting of the Technical Council of the Bureau in 1950.[60] Even after that, there arose a dispute between Rameev and Bazilevskii, who proposed developing a relay-type high-speed "computer" instead of a computer equipped with vacuum tubes. His proposal was also rejected.[61] Thus in 1950 the new direction of the Bureau to make a digital high-speed computer was confirmed. The Council of Ministers, which was probably informed of the new direction of SKB-245, added SKB-245 as another organization besides ITMiVT responsible for the development of digital computers in its Proclamation dated May 1951.[62] That ambiguous and compromising way of the Government made room for MMiP to behave avariciously for the monopoly in the development of all kinds of "mathematical machines." Now the Bureau entered into competition with the ITMiVT of the Academy of Sciences for a high-speed computer.

Parshin, Minister of MMiP, once invited Lavrent'ev to his Ministry. Sergei Vavilov, President of the Academy of Sciences, accompanied him. Parshin, saying, "I will make the machines. I have the ability. I called you, the Academy of Sciences, for help; you shall cover up the project with a scientific appearance, taking advantage of your scientific authority – by writing scientific papers when needed,"[63] made them aware of the Ministry's strong will for the monopoly of the production of computing machines.

The Bureau set about designing a universal electronic digital computer, which was later named "Strela" or "Arrow" (Figure 5.2). The timetable for this project was fixed by the Council of Ministers. It was supposed to be ready in

Figure 5.2 A general view of Strela-3 (RGAE F.8123, Op. 8, D.619. sheet 218).

the first quarter of 1953.[64] From a letter in October 1952 sent by the Director of the Bureau, we learn that the basic design of Strela was to include: (1) an external input mechanism consisting of a keyboard unit, a punch-card device, monitoring indicators, and entry device for magnetic tapes; (2) a magnetic-tape external storage unit; (3) a magnetic-drum internal storage unit; (4) an arithmetic operating device; (5) a controlling and signaling system; (6) an output mechanism consisting of a punch-card device and a printer-reader; and (7) an electric power supplier.[65]

As the Bureau was suffering from a lack of experts, the project faced long delays. In addition, the news about technological progress in this field in the United States often forced frequent changes in the design. The new requests were largely related: (1) the usage of electron-beam (cathode-ray) tubes as recording devices for the storage unit, and (2) the use of a germanium diode.[66] All these made the progress of the project slow. As late as 1953, some additional requests for the design were sent to the Bureau.[67] Thus, the completion of first Strela (Strela-1) was delayed until the end of 1953.[68] The first trial came only in the spring of the following year before a State Commission headed by Keldysh. It processed in about ten hours an integral-differential equation needing approximately 70 million calculating operations for its solution, which would have taken a mathematician about 100,000 days' work. The new additions to Strela-1 hardware included electron-beam tubes in the high-speed recording device, 125mm-wide strong magnetic tape for the external storage unit, and germanium diodes.[69] Strela-1 was able to process 2,000 arithmetic operations in a second. Its internal storage unit stored 1,023 words, with their numbers within a range of 10^{-19} to 10^{19} and the external storage unit stored about 200,000 words.[70] The State Commission, headed by Keldysh, concluded to recommend the practical use of Strela.[71]

Furthermore, they had taken several orders for Strelas by the date of the first trial of the machine. Among its clients were powerful military departments and institutions, such as the Ministry of Defense. Special Design Bureau No. 245 could not afford to cancel these orders. A year before the test of Strela, there had already been six plans to establish "Computing Centers": The Main Directorate of Artillery, the Naval Academy of the Ministry of Defense, ITMiVT, the Ministry of Aviation Industry (for the Central Aero-hydrodynamic Institute; *Tsentral'nyi Aerogidrodinamicheskii Institute*; TsAGI), and two other organizations, whose names were deleted from the documents. SKB-245 took charge of supplying Strelas for five of them, except for ITMiVT.[72] Including other orders, mainly for some higher education facilities, SKB-245 had taken totally eight orders by the time of the first test of Strela.[73] The official desire to set up "Computing Centers" then gained momentum. Taking advantage of this opportunity, the Bureau wanted to expand their organization into a complex employing as many as 1,300 personnel. In fact, there was a need for more experts, since there were several projects to bring to conclusion. What is more interesting, however, is that they showed a strong hostility toward the Academy of Sciences. They persuaded the Council of Ministers to have the Academy of Sciences transfer some of its personnel and equipment to the Bureau.[74] Thus, the Bureau was taking advantage of its strong connections with the military to promote its prestige and success, as well as to hide its failure. It appears that the Bureau even tried to get the military to blame the Academy of Sciences for the lack of success in computer technology in the Soviet Union. For example, in June 1955, Section No. 14 of the Scientific-Technical Council of the Ministry of Defense headed by Aksel' Berg (1893–1979), Deputy Minister of Defense, and, at the same time, Academician, adopted the "Recommendation," which was prepared by General Anatolii Kitov (1920–2005), Director of the Computing Center of the Ministry of Defense. He put blame on the Academy of Sciences for the serious failings and long delays that faced the completion of high-speed computers.[75] Taking this opportunity, the MMiP was soon appealing to the Council of Ministers to call the Academy to account for the matter.[76]

This success of Strelas was partly owing to the positive reputation of Keldysh's Commission, and partly owing to the delay of their rival, BESM, which resulted from a harassment made by MMiP. According to the extant records of MMiP, an experimental model of BESM machine achieved a speed of only 600–800 operations per second in 1953.[77] Monopolizing the electron-beam (cathode-ray) tubes for calculators in the Soviet Union, MMiP harassingly refused to supply them to ITMiVT, which could not help depending upon that ministry because of its lack of industrial basis.[78] When they used a storage unit consisting of electron-ray tubes, developed by the MMiP and the Ministry of Radio Industry in 1954 after long negotiations, BESM reached a speed as high as 7,000–8,000 operations per second.[79] It is also said that BESM-1 became one of the fastest computers in Europe (although not the world). According to the information of TASS (*Telegrafnoe agenstvo Sovetsrogo Soyuza*) on November 12, 1955, Dr. Dreier of Darmstadt Institute of Applied Mathematics admired BESM as the fastest machine in Europe.[80] Its memory was now built mainly with electron-beam tubes and ferrite magnetic cores in spite of the supersonic wave devices with which the BESM had been equipped before.[81]

Success of Strela was, however, greatly exaggerated, and does not reflect the reality. Anatolii Myamlin, an engineer who contributed to the development of Strela, confessed that the Strela installed in the Institute of Mathematics "worked very badly."[82] They apparently had to replace most of the hardware, except for the arithmetic operating device, only a year after the installation.[83] Mikhail Shura-Bura (1918–2008), a pioneer of Soviet computer science, for example, stated, "The machine very often failed and did not work properly."[84] What was worse, this machine was incompetent to solve mathematical tasks required for the study of nuclear fission. Aleksei Zabrodin (1933–2008) remembered that Yurii Khariton (1904–1996), a leading scientific expert of the Soviet nuclear weapon project, visited Keldysh to talk about the solutions of complicated quadratic equations. Zabrodin witnessed that Strela was not able to solve these equations.[85]

Then, why was Strala highly evaluated by the State Commission? Strela was visually a nice-looking machine to display. It had hundreds of lamps, most of them blinking. A witness described it as a "beautiful machine in a beautiful hall."[86] Keldysh made use of that beautiful physical appearance of the machine to hide its failings. Now Keldysh, who had been on the side of Lavrent'ev and his Institute, had to pretend to be satisfied with the success of Strela, because the initial failure of the computer development would have been a political disaster. This was perhaps a matter of prestige for him and for his institution. He, therefore, worked very hard to conceal the relative failure of the development of the first Soviet computer from the eyes of the central planners.[87] This ambiguous attitude eventually allowed the political, but momentary, success of Strela.

A crucial step came when, in July 1955, Igor' Kurchatov (1903–1960) and Sergei Sobolev (1908–1989), a leading mathematician, asked MMiP to provide BESM computers for their Laboratory of Measurement Instruments, the successor of "Laboratory No. 2," the first research center for atomic bomb development in the Soviet Union, and now the Kurchatov Institute.[88] In practice, this meant the end of the Strela project and as a result the MMiP canceled the Strela project.[89]

Notes

1 Rossiiskii gosudarstvennyi arkhiv sotsial'no-politicheskoi istorii (RGASPI): Fond (F.) 17, Opis' (Op.) 125, Delo (D.) 447, 29, 30.
2 Arkhiv Rossiskoi Akademii nauk (ARAN), Fond (F.) 1559, Opis' (Op.) 1, Delo (D.) 3, 1.
3 ARAN, F. 395, Op. 1–50g, D. 1504-2-613, 87.
4 RGASPI, F. 17, Op. 132, D. 36, 29: Bruevich was teaching at the Military Aviation Academy from 1929–1961. He was a Lieutenant General during World War II (N. Krementsov, *Stalinist Science*. Princeton and Oxford, Princeton University Press, 1997, 98, 300).
5 ARAN, F. 1559, Op. 1, D. 3, 3, 4.
6 Cited in G.D. Crowe and S.E. Goodman, "S.A. Levedev and the Birth of Soviet Computing." *IEEE Annals of the History of Computing*, Vol. 16, No. 1, Spring 1994, 14.
7 RGASPI, F. 17, Op. 132, D. 36, 48–49.
8 Rossiiskii gosudarstvennyi arkhiv ekonomiki (RGAE), Fond (F.) 8123, Opis' (Op.) 8, Delo (D.) 308, 31.
9 RGAE, F. 8123, Op. 8, D. 402, 1.
10 RGAE, F. 8123, Op. 8, D. 524, 27.
11 RGAE, F. 8123, Op. 8, D. 523, 11.
12 RGAE, F. 8123, Op. 8, D. 308, 29–31.

13 RGAE, F. 8123, Op. 8, D. 482, 85: The major clients at that time were the Ministry of Armed Forces (later – Ministry of Defense; 1,091,600 rubles), Moscow State University (3,201,600 rubles), and Factory No. 293 of the Ministry of Aviation Industry (623,300 rubles).

14 Here let us examine what kind of mathematical machines they developed and who the clients were. In 1951, the Bureau developed a small-sized analyzer, called "Integral I," equipped with four integrators (RGAE, F. 8123, Op. 8, D. 402, 1). In the same year, the Bureau began to conduct research on a calculator for discrete differential equations, and a special-purpose analogue calculator called "EV-80–3." The latter was demanded by Pavel Zernov, Director of the 'Construction Bureau No. 11', the Russian research and development center for nuclear weapons known under the code of "Arzamas-16" (RGAE, F. 8123, Op. 8, D. 481, 43, 44). Consequently, in 1952, the electrical integrator called "Eli-6," and the electronic integrators, IPT-4, IPT-5, and IPT-6 were developed by the Bureau, and began to be manufactured in the Penza SAM (RGAE, F. 8123, Op. 8, D. 523, 11). The following year, the Bureau built the automatic high-speed computing machines S-1 and S-2, specialized electronic calculators for general differential equations, "Polyot" (or "flight," a computing machine specialized for testing automatic piloting systems of aircraft), "Operator" (an analogue computer for modeling with changeable variables), "Integral VI (with 6 integrators)," and an analogue calculator for discrete differential equations (ibid., 94–109). It ventured to develop a high-performance differential analyzer equipped with 24 integrators, called MDA, a high-performance electrical integrator, a high-speed relay calculator, an integrator called IPT-9, and two sorts of high-resistance germanium diodes (ibid., 153–156). The Bureau also received contracts for specialized calculators for general differential equations, STsM, "Pogoda" ("Weather"; a calculator for a series of double-entry products concerned with weather forecasting), "Kristall" ("Crystal," a computing machine used for X-ray analysis), and various high-performance relay-type calculators (ibid., 161–163). Even in the second half of the 1950s, the Bureau continued to spend its energy on the production of special-purpose calculating and computing machines, such as differential analyzers, many of which were used for trajectory calculations. The Bureau planned to undertake to develop four new machines for the Main Directorate of Artillery of the Ministry of Defense in the period from 1955 to 1957: "Korund" ("Corundum," a calculator specialized for the trajectory characterizing of the atmosphere), "Priyut" ("Hiding place," a calculator designed for geological surveys), "Granit" ("Granite," a calculator specialized for verifying accounts of correlative functions.), and "Udar" ("Blow," a calculator used for integrations of probability of strikes). The Bureau also planned to build, probably for the Main Directorate of Artillery; a SKB-245 document indicates that these three projects were to be carried out in order to fulfill the technical demands of the user who was to use these machines for the sake of the Main Directorate of Artillery of the Ministry of Defense according to the agreements (RGAE, F. 8123, Op. 8, D. 629, 185, 186, 188), three other new devices: "Agat" ("Agate," a calculator specialized for control of ground artillery guns against ground or flying targets), "Poloz" ("Slip," or "roller," a calculating system for the firing mechanism of ground artillery guns), and "Poligon" ("Firing range," a calculator to process calculations in realistic conditions). In addition, they planned to build a machine called "Mramor" ("Marble," a correlative analyzer for super low frequency bands), for Division No. 27177 of the Main Directorate of Naval Shipyards of the Ministry of Defense (ibid., 181–188). In 1955, the Central Institute of Aeronautical Hydrodynamics (*Tsentral'nyi Aero-Gidrodinamicheskii Institut*; TsAGI) recommended a calculator with an analogue calculating mechanism for the complex simulation of jet plane flights, MiG-19p and Tu-16 (RGAE, F. 8123, Op. 11, Edinits khraneniya 11, 181–188).

15 The Energetics Institute of the Academy of Sciences, for example, had been striving to develop differential analyzers equipped with six integrators since 1936 (RGAE, F. 8123, Op. 8, D. 524, 19). The first device that it designed was manufactured in the Sestretsk Armory between 1938 and 1939, and was in use at the Institute even in the postwar period (ibid., 23). The same institute had also been endeavoring for ten years to design electronic integrators for differential equations. From 1948 to 1952, among the scientific pursuits of the Academy of Sciences was a small-sized electronic analogue computing machine (ibid., 19).
16 Rossiiskaya Akademiya nauk, Ordena Lenina Sibirskoe otdelenie, *Vek Lavrent'eva*. Novosibirsk, Izdatel'stvo SO RAN, 2000g, 54.
17 RGASPI, F. 17, Op. 132, D. 36, 48, 49.
18 A.V. Zabrodin (Ed.), *M.V. Keldysh: Tvorcheskii Portret po Vospominaniyam Sovremennikov*. Moscow, Nauka, 2002g, 344: The calculator "Mercedes" was invented by Ch. Hamann in Zella-Mehris in 1905, and was produced in eastern Germany (the author thanks Prof. M. Ludwig for this reference). Unfortunately no documents showing the reality of the young girls who were engaged in the calculations in the Soviet Union at that time could be found. Since the publication of the article by Jennifer S. Light, "When Computers Were Women" (*Technology and Culture*, Vol. 40, No. 3, July 1999, 455–483), the role of young girls in the first days of the computing age is getting more and more attention.
19 A.N. Myamlin, "M.V. Keldysh i Vychslichel'naya Tekhnika." in Zabrodin (Ed.), *op. cit.*, in note 18. 343.
20 ARAN, F. 1559, Op. 1, D. 15, 15.
21 S. Gerovitch, *From Newspeak to Cyberspeak: A History of Soviet Cybernetics*. Cambridge, MA, London, The MIT Press, 2002, 132, 133.
22 *Vek Lavrent'eva*, *op. cit.*, in note 16. 77.
23 As for the reconsideration of the meaning of these "Scientific Discussions" on the basis of the newly opened documents, see A.B. Kozhevnikov, *Stalin's Great Science: The Times and Adventures of Soviet Physicists*. Imperial College Press, 2004, 186–216. Especially concerning the "Philosophical Discussion," the very beginning of the whole process, see, Yu.I. Krivonosov, "Srazhenie na filosovskom fronte: Filosovskaya diskussiya 1947 goda – prolog ideologicheskogo pogroma nauki." *Voprosy Istorii Estestvoznaniya i Tekhniki*, No. 3, 1997, 63–86, and V.D. Esakov, "K istorii filosovskoi diskussii 1947 goda." *Voprosy Filosofii*, No. 2, 1993, 83–106. For example, concerning the influence of a planned discussion upon a field of science (physics), see, V.P. Vizgin, "Yadernyi shchit v 'Tridtsatiletnei voine' fizikov s nevezhestvennoi kritikoi sovremennykh fizicheskikh teorii." *Uspekhi Fizicheskikh Nauk*, Vol. 169, No. 12, December 1999, 1364–1389 (especially, 1371–1383).
24 *Vek Lavrent'eva*, *op. cit.*, in note 16. 58.
25 Crow, Goodman, *op. cit.*, in note 6. 4–8.
26 *Vek Lavrent'eva*, *op. cit.*, in note 16. 59, 60: Later Khrushchev's patronage for Lavrent'ev became well known. As late as 1962, Lavrent'ev did not forget to express his thanks to Khrushchev in his speech. He says, "Although the worldwide center of mathematics has moved to the United States owing to the immigration of many mathematicians (John von Neumann and others), and many great mathematicians have been mobilized to the physics field in the Soviet Union, we have been able to gather a sufficiently talented collective. N.S. Khrushchev gave us the powerful and organizational support." (Nauchinyi arkhiv Sibirskogo otdeleniya Rossiiskoi Akademii nauk. F. 27, Op. 1, D. 76, 4).
27 *Vek Lavrent'eva . . . op. cit.*, in note 16. 60.
28 Citing the minutes of the Party's Politburo meetings and other documents, Gerovitch affirms that Khrushchev took some actions in favor of Lavrent'ev in the Party's top echelons (Gerovitch, *op. cit.*, in note 21. 136).

29 RGASPI, F. 17, Op. 132, D. 36, 34–44.
30 RGASPI, F. 17, Op. 132, D. 36, 47.
31 ARAN, F. 395, Op. 1–49g, D. 1503-62-613/sheets 149–172: In this "reply," they say "a series of the machines, based upon analogue principles, are younger. They have only appeared in the 20th century. The basic theories and the first example in the world of a differentiate analyzer were developed by the late academician, A.I. Krylov (ibid., 156) . . . . The numerical machines cannot be massive computers for the solution of analytical subjects (ibid., 158) . . . . ENIAC has a large number of vacuum tubes. Up to 70% of these tubes may, however, come to a standstill, which is somehow undesirable in the results and needs an inspection by special testing . . . . We must not repeat ENIAC (ibid., 160). Research considering the problems of digital, high-speed, automatic machines began in ITNiVT in the fall of 1948 with the help of the correspondent member of the Academy of Sciences, I.S. Bruk . . . . It became known that the subjects of the construction of such machines are so complicated that a special resolution is needed" (ibid., 162).
32 Ibid., 2–4.
33 Ibid., 75.
34 ARAN, F. 1559, Op. 1, D. 6, 11.
35 Ibid., 51.
36 ARAN, F. 2, Op. 3-a, D. 109, Microfish.
37 Ibid.: He kept this post until his removal from ITMiVT in the fall of 1950.
38 ARAN, F. 395, Op. 1–50g, D. 1504-60-613, 3.
39 Although his proposal was eventually accepted in the Bureau meeting of May 23, some members showed some hesitation: Hearing his request asking for the appointments of new members of the Scientific Council, Boris Vvedenskii, Academician – Secretary of the Department, said, "This proposal must be argued in the presence of Lavrent'ev." Ivan Oding added to this remark, "Yes. This is a significant turning point." S.I. Kostin, however, said, "I think that this proposal should be approved. As for that problem, we've already come to a conclusion." A famous metallurgist, Ivan Bardin (1883–1960), expressed his approval to Kostin. Finally Vvedenskii declared the end of the deliberation with a shout, "Adopted!" (ARAN, F. 395, Op. 1–50g, D. 1504-65-613, 31). Moreover, inside the Institute Lavrent'ev met with strong resistance; the Scientific Secretary (Dzhangil N-M-O. Shakhsuvarov), who was removed from office by Lavrent'ev, accused him in court of unfair dismissal and won the trial in the lower courts (*Vek Lavrent'eva, . . . op. cit.*, in note 16. 59, 60).
40 ARAN, F. 395, Op. 1–50g, D. 1504-69-613, 59, 63, 66, 68: Bruevich said, "Two years ago, I submitted the leadership a request for my resignation from the post of Academician-Secretary of the Academy of Sciences (more precisely, Academician-Secretary of the Presidium of the Academy – *Ichikawa*). Then I was motivated by two emotions. Owing to the heavy burden of this leading post, I could not engage myself in the scientific activities related to my own specialty for fourteen years. After my long illness, after a series of the happenings, one year later, my resignation was accepted. Through the good offices of those who took my age of fifty-four years and my illness into consideration, I was appointed the Director of the Institute of Precision Mechanics and Computing Technique and, at the same time, Chief of a department of the Institute. And, now, I cannot help referring to my right to carry out scientific research activities in the fields in which I have been engaged for a long time. I have been engaged in two fields – precision of machinery and computing techniques. . . . I had carried out the task of computing continuous functions for a very long time. . . . Although I was engaged in this task, i.e. the development of differentiate analyzers, for many years, why is it that we cannot find my surname in the Five Year Plan? . . . This (the direction which Lavrent'ev was taking – *author*) means the adoption of a measure of exiling Bruevich. This is not a Soviet attitude towards the scientist. I think that I need to

make a proper protest" (ibid., 59, 63). Contrary to him, Lavrent'ev brought forth a counterargument to Bruevich's remark; "I won't refer to any personal matter of anyone without the direction of Boris Alekseivich (Vvedenskii). Not only Nikolai Grigorievich (Bruevich) is devoting himself to the task of differentiate analyzers. I don't regard Nikolai Grigorievich as the only specialist in this field" (ibid., 66). Bruevich tried again to make a counterargument against Lavrent'ev, saying, "I can understand that the Department of Technological Sciences will argue about the matter of personality independently. Second. Appointing the leader and removing the leader are quite different things." He tried to continue, "The matter would be different, if I had committed any mistake in leadership." No sooner had he said that, a Bureau member, Anatolii Blagonravov (1894–1975), interrupted, saying, "I propose that the plans should be adopted." Thus the argument was stopped with Vvedenskii shouting, "Adopted!" (ibid., 68).

41 ARAN, F. 1559, Op. 1, D. 15, 1–82.
42 Ibid., 28.
43 Ibid.
44 Ibid., 76–81.
45 RGASPI, F. 17, Op. 133, D. 174, 71; ARAN, F. 1559, Op. 1, D. 14, 87.
46 RGASPI, F. 17, Op. 132, D. 354, 185, 186.
47 Ibid., 187.
48 See, for example, his letter to Vyacheslav Molotov (1890–1986), Vice-Chairman of the Council of People's Commissars, dated August 7, 1943 (RGASPI, F. 82, Op. 2, D. 930, 49–56). He, in the capacity of Academician-Secretary of the Presidium of the Academy, revealed some calumnious information about some candidates for membership of the Academy. For example, he says, "L.D. Landau is politically ineligible," "B.G. Kuznetsov is zero (totally useless – *Ichikawa*)," "It is undesirable that two persons of German origin are nominated for the Presidium while we are fighting with Germany," and so on. Also in the postwar days, he sent a letter to the Party's Central Committee Secretary, Aleksei A. Kuznetsov (ibid., 141–148) in order to inform him of the political careers and the nationalities of the scientific workers of the Institute of History, the Institute of Physics, and the Department of Biological Sciences of the Academy of Sciences. He also made recommendations about a series of personnel matters.
49 Immediately after the dismissal of Bruevich from ITMiVT, at the suggestion of Lavrentii Beriya (1899–1953), Minister of Home Affairs, the Academy of Sciences carried out a survey on the activities for observation of secrecy in the Department of Precision Mechanics and Machines of the Institute of Mechanical Engineering. They found Bruevich and other associates leaving classified experiment data on the accuracy tests of the bombing-meters, which were carried out in Baku in 1944–1945 at the demands of the Navy, open on their desks. Consequently five of them, including Bruevich himself, received heavy official reprimands for a lack of wariness, and senior associate Kobrinskii also received a dismissal notice (RGASPI, F. 17, Op. 133, D. 174, 64–66).
50 RGASPI, F. 17, Op. 132, D. 354, 188–191.
51 RGAE, F. 8123, Op. 8, D. 524, 20.
52 Ibid.
53 Crow, Goodman, *op. cit.*, in note 6. 8, 9.
54 RGAE, F. 8123, Op. 8, D. 623, 113, 114.
55 RGAE, F. 8123, Op. 8, D. 623, 89–91.
56 B.N. Malinovskii, *Istoriya Vychislitel'noi Tekhniki v Litsakh*. Kiev, 1995. Electronic Version, Chap. 6, http://lib.ru/MEMUARY/MALINOWSKIJ/7.htm, 6.
57 For his contribution, see, Malinovskii, *op. cit.*, in note 56. Chap. 5, http://lib.ru/MEMUARY/MALINOWSKIJ/6.htm.
58 Malinovskii, *op. cit.*, in note 56. Chap. 6, http://lib.ru/MEMUARY/MALINOWSKIJ/7.htm, 17.

59 Malinovskii, *op. cit.*, in note 56. Chap. 2, http://lib.ru/MEMUARY/MALINOWSKIJ/3.htm, 2: D. Kipyatkov, Chief of Third Division of the State Planning Committee, said in his report dated February 21, 1951, "Industry has a serious difficulty in gaining the necessary amount of vacuum-tubes" (RGAE, F. 4372, Op. 98, D. 968, 25). It was planned that a billion rubles would be invested in vacuum-tube manufacture from 1951 to 1955. The main purpose of this investment was, however, to enlarge the industrial basis of Soviet radar-building (ibid., 28).
60 Malinovskii, *op. cit.*, in note 56. Chap. 2, http://lib.ru/MEMUARY/MALINOWSKIJ/3.htm, 4.
61 Malinovskii, *op. cit.*, in note 56. Chap. 6, http://lib.ru/MEMUARY/MALINOWSKIJ/7.htm, 19.
62 RGAE, F. 8123, Op. 8, D. 481, 89.
63 *Vek Lavrent'eva, . . . op. cit.*, in note 16. 60.
64 RGAE, F. 8123, Op. 8, D. 481, 89: It is noteworthy that the initial proposal by MMiP for the financial plan for this machine was denied by the Ministry of Finance and State Planning Committee and the total sum of expenditure for the development was reduced to only 200,000 rubles (RGAE, F. 8123, Op. 8, D. 345, 81).
65 RGAE, F. 8123, Op. 8, D. 481, 89.
66 RGAE, F. 8123, Op. 8, D. 402, 1, 2.
67 RGAE, F. 8123, Op. 8, D. 523, 84.
68 As Scientific Research Institute No. 160 of Ministry of Industry for Communication Means, which was asked to develop the electron-beam tube, rejected signing a contract on account of their "lack of experience," SKB-245 requested the Vice-Chairman of Council of Ministers, Vyacheslav Malyshev (1902–1957), to make that ministry manufacture the tubes as early as they could (RGAE, F. 8123, Op. 8, D. 481, 102). Because of this and other reasons, the design and manufacturing of "Strela" by SKB-245 were delayed to some extent (RGAE, F. 8123, Op. 8, D. 402, 2).
69 RGAE, F. 8123, Op. 8, D. 560, 200–202, 205–208.
70 Ibid., 205–208.
71 Ibid.
72 RGAE, F. 8123, Op. 8, D. 524, 25.
73 RGAE, F. 8123, Op. 8, D. 560, 192.
74 Ibid., 263, 295.
75 RGAE, F. 8123, Op. 8, D. 623, sheet 253: For the content of the discussion in this section, see RGAE, F. 8123, Op. 8, D. 630, 18–24.
76 RGAE, F. 8123, Op. 8, D. 623, 138–142.
77 Ibid., 138.
78 *Vek Lavrent'eva, . . . op. cit.* 59, 60.
79 RGAE, F. 8123, Op. 8, D. 623, 112.
80 I.M. Makarova et al. (Ed.), *Istoriya Informatiki v Rossii: Uchenye i Ikh Shkoly*. Moscow, Nauka, 2003, 74.
81 S.P. Prokhorov, "Computers in Russia: Science, Education and Industry." *IEEE Annals of the History of Computing*, Vol. 21, No. 3, July–September 1999, 4.
82 Myamlin, Pod. otvet. red. Zabrodina, *op. cit.*, in note 18. 343.
83 Ibid.
84 M.R. Shura-Bura, "Moi Keldysh", Pod. otvet. red. Zabrodina, *op. cit.*, in note 18. 359.
85 A.V. Zabrodin, "V nachale bol'shogo puti", ibid., 370.
86 Shura-Bura, *op. cit.*, in note 84. 358, 359.
87 Ibid., 359, 360.
88 RGAE, F. 8123, Op. 8, D. 623, 271, 272.
89 Ibid., 261.

Part III

"Hope" and "anxiety"

Soviet science at the dawn of the nuclear age

6 Obninsk, 1955

The world's first nuclear power plant and "the atomic diplomacy" by Soviet scientists

The first nuclear power plant in the world went into operation at Obninsk, a southern suburb of Moscow, in July 1954. One year later, the Soviet Academy of Sciences held a large-scale international conference, the Session of the Academy of Sciences of the USSR on the Peaceful Use of Atomic Energy, as an international stage to demonstrate their scientific achievements in nuclear science and technology. Soviet scientists subsequently challenged the United States' nuclear advancement at the First United Nations International Conference on the Peaceful Use of Atomic Energy at Geneva. Could the Soviet scientists succeed in gaining international acceptance and prestige? Could they gain self-confidence?

1 The unit "AM": a submarine reactor out of water

The decision to develop an atomic engine for a submarine for the first time in the Soviet Union was made at a meeting of the First Main Directorate held on February 11, 1950,[1] which was then in charge of the Soviet nuclear project. They could not, however, gain any essential technical information from the U.S. or other Western countries due to their very rigid control and restriction on information distribution in the midst of the Cold War. In such circumstances, the chief reactor designer, Nikolai Dollezhal' (1899–2000), worked with other designers on a plan to minimize a graphite reactor in terms of the scale and weight and load it onto a submarine. Dollezhal' explains the reason for their choice of a "solid" reactor as follows: They aimed "to exclude the phenomena which might be caused by a 'liquidized' reactor."[2] The graphite reactor that they aimed to develop was codenamed the unit "AM."[3]

In retrospect, such an attempt was destined to fail. The policy change from the utilization of a graphite reactor for a submarine engine to that of a light-water reactor (and a liquid metal cooled fast neutron reactor) was ultimately confirmed in the Proclamation of the USSR Council of Ministers dated July 28, 1953.[4] It seems, however, that the infeasibility of their initial attempt had become clear long before that proclamation. As early as March 28, 1950, Major General and First Deputy Chief of the First Main Directorate Avraamii Zavenyagin (1901–1956) and Academic Secretary of the Scientific and Technical Council of the

Directorate Boris Pozdnyakov (1903–1979) made a proposal under their joint signature to the Directorate to settle on a plan to utilize atomic energy for the sake of the national economy. The fact that very detailed numerical indices of the design of a new reactor for this purpose were enclosed with the proposal[5] suggests that definite progress had been made on the design works for this type of nuclear reactor. Their proposal was accepted by the government. The Proclamation of the USSR Council of Ministers No. 2030–788 dated May 16, 1950, titled "On the scientific research, design and experimental activities concerning the utilization of atomic energy for peaceful purpose," combined with its supplements issued on July 8, 1950, marked a new age of atomic energy in the Soviet Union.[6]

1.1 "AM" and the political leadership

On November 10, 1949, approximately only two months after the first explosion test of the Soviet atomic bomb, the Soviet representative to the United Nations, Andrei Vyshinskii (1883–1954), referred to the non-military use of atomic energy for the first time in the international arena of the Fourth General Assembly of the United Nations, saying,

> In the Soviet Union we utilize atomic energy not for increasing the stock of atomic bombs. . . . We are utilizing it for the sake of our interests of economic and managerial character. We will make good use of atomic energy in order to crush the mountains, change the river streams, irrigate wasteland and, thus, open our way for a new life in the places where mankind had hardly stepped into.[7]

This speech, which was the first official remark made by the Soviet political leadership on the peaceful use of atomic energy, was in part anti-American propaganda and in part a self-serving justification for the Soviet atomic bomb, the weapon which the Soviet Union had earlier denounced as "the most serious international crime against the human being" at a meeting of the UN Atomic Energy Committee on June 19, 1946, where it proposed the total prohibition of making and using any atomic bomb.[8]

Therefore, it seems that the Soviet political authority, rather than condemning the reactor designers, welcomed the unforeseen emergence of a plan to use the nuclear reactor for a "peace purpose," which by then had already become useless for the initial purpose of building a submarine engine. On October 5, 1952, a year and two months earlier than the "Atoms for Peace" address made by U.S. President Dwight Eisenhower (1890–1969), a Politburo member, Georgii Malen'kov (1902–1988), spoke highly of the peaceful use of atomic energy at the first day of the meeting of the Nineteenth Convention of the All-Union Communist Party (Bol'sheviki):

> The discovery of methods to produce atomic energy is the most important progress in Soviet science these days. Our science and technology overthrew the monopolizing status of the United States in this area, causing

some serious damage to the war arsonist that tries to intimidate and threaten other peoples by taking advantage of their monopoly of the secrets of atomic energy production and atomic weapons. The Soviet state that has learned to produce atomic energy is actually very interested in the utilization of this new energy for peaceful purposes and for the people.[9]

1.2 "AM" and scientists

It must also be taken into consideration that some leading Soviet scientists had already urged the political authorities to develop non-military use of nuclear power since the days immediately after the end of World War II.

The first Soviet scientist who appealed to the government agencies to develop nuclear power for peaceful use was a famous physicist, Pyotr Kapitsa (1894–1984. A Nobel Prize laureate in Physics in 1978). His proposal is assumed to have been made as early as in the autumn of 1945. On October 26, 1945, it was announced at a meeting of the Technical Council (later renamed the Scientific and Technical Council and placed under the First Main Directorate) that a proposal about "the Peaceful Atom" made by Kapitsa was to be considered at the "Special Commission under the State Defense Committee."[10] This commission was then headed by a Deputy Primer and Politburo member, Lavrentii Beriya (1899–1953), who supervised the Soviet nuclear development project. It was later dissolved

Figure 6.1 Upper shielding of the reactor at the Obninsk atomic power station.

Source: State Scientific Centre of the Russian Federation, Institute of Physics and Power Engineering, *The World's First Nuclear Power Plant: Documents, Articles, Memoirs, Photographs.* Obninsk, 2014. 185.

Figure 6.2 An outward appearance of an "atomic reactor."

Source: K.A. Gladkov, Energy of Atom. Detigiz, 1958. 165. This illustration does not look like the "AM" at all.

in February 1946 and its function was transferred to the Council of Ministers. Determined, Kapitsa also sent a letter offering his proposal to another Deputy Primer and Politburo member, Vyacheslav Molotov (1890–1986), on December 18, 1945.[11] Three months later, on February 12, 1946, the Soviet nuclear project leader, Igor Kurchatov (1903–1960), who was later called the "Atomic Tsar," sent a report addressed to Stalin suggesting a rosy near future of the application of atomic energy to technology, chemistry, biology, and medicine.[12] In addition, on April 22, 1946, President of the Soviet Academy of Sciences Sergei Vavilov (1891–1951) sent Beriya (through his deputy, Vasilii Makhnyov – 1904–1965 –) a "memorandum on the organization of research activities related to atomic nuclear energy in various fields of science and technology."[13]

What is the most noteworthy here is that Kruchatov placed a very strong emphasis on the desirability and adequacy of control over the project by the Academy of Sciences. He said the following in his report dated February 12, 1946:

> In accordance with this [initiation of the peaceful atom project – *Ichikawa*], the USSR Academy of Sciences should be delegated authority to organize the studies on the application of nuclear power and the radioactive materials in technology, chemistry, biology and medicine, as an issue of primary significance and to mobilize the scientists who have not worked on nuclear power yet.[14]

In the Soviet era, even during its Stalinist period, the Academy of Sciences kept its autonomy to some extent, often carried out negotiations with the political authorities, and sometimes provided scientists with "shelter" against political power intervening in science.[15] In the atomic project, however, a governmental agency affiliated with the USSR Council of Ministers, namely the First Main Directorate, directly organized research and development activities, mobilizing a large number of scientists. That was a serious encroachment upon the organization of the Academy; indeed, the Leningrad Physical Engineering Institute – from which many research fellows, including Kurchatov, were mobilized – fell into functional insufficiency.[16] The Academy of Sciences could not remain unconcerned with further encroachment upon its organization. In addition to this, on December 21, 1945, Kapitsa was removed from his post in the "Special Commission," in the Technical Council, and in other organizations (he only retained his full membership in the Academy of Sciences) due to a feud with Beria mainly over the freedom of publication of research results. Kapitsa was forced to confine himself in his own house (dacha).[17] This incident made many leading scientists afraid of possible interferences from political authorities. Therefore, it was desirable for scientists to try and put a big scientific project of the state under their own direct control as a way to protect themselves from the interference of political power. On the other side, considering the time and effort required to manage a big new project that was lacking urgency, the Party and government preferred entrusting such a burdensome project that only had ideological significance to an existing organization, i.e., the Academy of Sciences. Thus lobbying by scientists bore fruit: The division of labor between the Academy of Sciences in charge of "the peaceful use" of atomic energy and the government in charge of nuclear weapon development was formalized in the Proclamation of the USSR Council of Ministers dated December 16, 1946, titled "On the development of scientific research activities on the atomic nuclear study and the utilization of nuclear energy in technology, medicine and biology."[18] The main parts of Sergey Vavilov's proposal were adopted in this proclamation.

1.3 The real "AM"

The Obninsk Atomic Power Station, whose reactor reached the critical state on June 27, 1954, was the first commercial nuclear power plant in the whole world and thus attracted worldwide attention. During the decade following its opening, the plant accepted about 39,000 visitors, including approximately 7,200 foreigners from 65 countries.[19]

In total, 157 holes, each of which had a diameter of 65 mm, were dug vertically with each pitch arranged so as to form an equilateral triangle of 120 mm on each side on the surface of the graphite foundation molded into a cylindrical form of 3,000 mm in diameter and 4,600 mm in height. The reactor required at least 550 kg of fuel, including 27.7 kg of uranium-235; uranium enriched to 5% was used as the fuel.[20] The reactor went critical in 1954, but reached its full capacity only in 1960. That reactor was equipped with a secondary water circle in the cooling system, which reduced thermal efficiency. The control rods filled with boron-carbide in stainless steel tubes were so insufficient in heat resistance that it also became necessary to induce cooling water even into the control rods. Without the cooling water, due to the low airtightness of the control rods, air would have entered the space between the tube and the neutron-absorbing material, and then some tubes would have become so heat-expanded due to high temperatures that they would have been damaged. Such a situation also occurred with the fuel rods. The engineers had to pay close attention to the airtightness of the fuel rods. Nevertheless, a fuel rod rupture accident occurred in 1959.[21]

Although that nuclear reactor had a heat-generating capacity of 30 megawatts, its electrical power generating capacity was only 5 megawatts. Therefore, as Lev Kochetkov (Figure 6.3) later testified in the interview mentioned in note 6, this reactor was an experimental rather than a reactor intended for commercial use.[22] However, the reactor had enough appeal to win the praise of many political leaders of the world, especially from Asia; Obninsk' visitors included Jawaharlal Nehru (1889–1964; Figure 6.4), the first Prime Minister of independent India, along with his daughter Indira Gandhi (1917–1984), Sukarno (1901–1970), the President of Indonesia, Ho Chi Min (1890–1969), the President of the Democratic Republic of Vietnam, Kim Il Sung (1912–1994) from North Korea, and others.[23]

2 The Session of the USSR Academy of Sciences on the Peaceful Use of Atomic Energy

2.1 *Hasty invitations*

After the death of Stalin in March 1953, thanks to a slight and passing thawing of tension in the Cold War, Soviet scientists' international ties began to rapidly and significantly expand. While 114 Soviet scientists in total went abroad as members of delegations or in individual capacities within the framework of the USSR Academy of Sciences in 1953, the number of overseas visitors increased to 175 in 1954 and then to 481 in 1955. During the same period, the number of foreign scientists who visited the Soviet Union at the invitation of the Academy of Sciences increased almost four times (from 93 to 362).[24]

In such an exciting circumstance, the USSR Academy of Sciences held the Session on the Peaceful Use of Atomic Energy on July 1–5, 1955, only three months after a conference on theoretical physics was held.[25] The Academy sent invitations to influential scientific institutions of various countries, including the Royal

Figure 6.3 Lev Kochetkov (left) explaining about the manipulation of the Obninsk Atomic Power Station to a TV crew from Japan.

Source: The collection of the author.

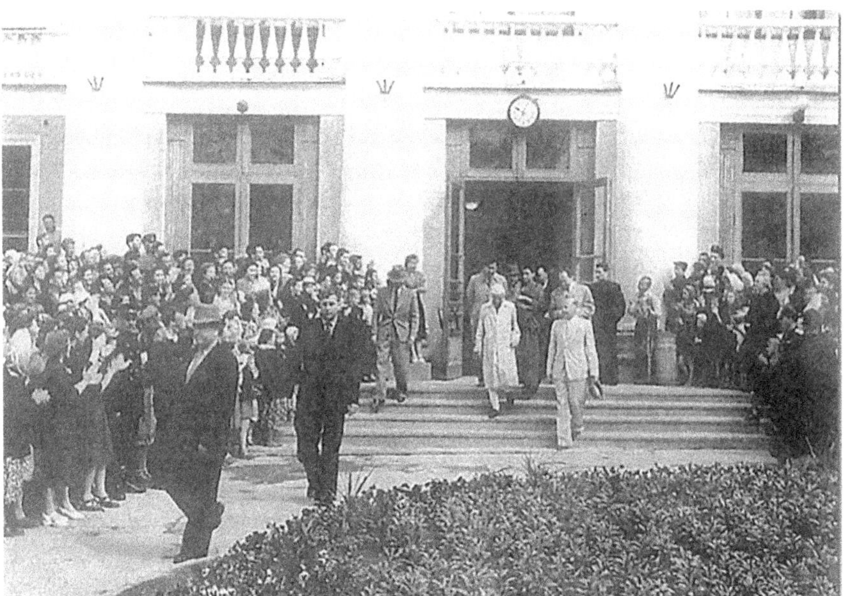

Figure 6.4 Jawaharlal Nehru on visit to Obninsk.

Source: State Scientific Centre of the Russian Federation, Institute of Physics and Power Engineering, *The World's First Nuclear Power Plant: Documents, Articles, Memoirs, Photographs*. Obninsk, 2014. 190–191.

Society, Académie des Sciences, Max-Planck-Gesellschaft, the Science Council of Japan (Nihon Gakujutsu Kaigi), and others mainly through the formal diplomatic channels. They also sent the invitations directly to leading scientists, many of whom were Nobel laureates, such as Patrick Blackett (1897–1974), John Cockcroft (1897–1967), Niels Bohr (1885–1962) and his son, Aage (1922–2009), Werner von Heisenberg (1901–1976), Chandrasekhara Venkata Raman (1888–1970), Robert Oppenheimer (1904–1967), Harold Urey (1893–1981), Ernest Laurence (1901–1958), Fransis Perrin (1901–1992), Lev Kovarsky, Sakata Shôichi (1911–1970), Yukawa Hideki (1907–1981), and others.[26] Almost all of them, however, could not accept the invitations due to insufficient time for preparation. The Royal Society replied to the Soviet Union Ambassador to the United Kingdom on June 27, 1955, as follows:

> Dear Ambassador, It is with very great regret that I have to let you know that the time at our disposal has proved too short for us to arrange a suitable small delegation to the Conference on the Peaceful Uses of Atomic Energy which is about to start in Moscow on the 1st July.[27]

The National Research Council of Canada answered in the name of President E.W.R. Steacie on June 30, 1955, addressing their response to Aleksandr Nesmeyanov (1898–1980), then President of the Soviet Academy of Sciences: "I wish to express my sincere regret that for the reasons given it is impossible for us to accept your very kind invitation to participate in the Session of the Academy on the Peaceful Uses of Atomic Energy."[28] Ernest Lawrence replied in a letter dated June 22, 1955, to Norair Sisakyan (1907–1966) – who was then a correspondent member of the Academy of Sciences playing an active role in the field of international relations – as follows: "As I have not been able to rearrange my schedule at this late date in order to come over for the sessions, I am all the more regretful in the view of the splendid program."[29] Heisenberg replied, "I am sorry that other tasks of mine prevent me from taking part in the Session."[30] The Science Council of Japan received a telegram from Nesmeyanov on June 10, 1955, proposing dispatch of a representative to the Session, who would be accommodated as a guest for two weeks. Yoshio Fujioka (1903–1976), who eventually participated in the Session, looked back to those days and reflected thusly:

> In order to go to Moscow until July 1, tremendous procedures and the shortage of funds for travels made it difficult for the Science Council to dispatch any representative to the Session. Fortunately, however, I had been already scheduled to attend the Board of the International Physics Conference which was to be held in Switzerland in mid-July. In addition to this, it was pure chance that I was Chair of Committee on Atomic Energy of the Council. Therefore, they [some members of the Science Council of Japan – *Ichikawa*] said that it was better for me to take part in the Session in Moscow on my way to Switzerland.[31]

The expansion of international activities by the Academy of Sciences, however, was not matched by the improvement of organizational work in this area. A tremendous amount of bureaucratic tasks, lack of clear organization, and inexperience as a whole led to delay and inopportuneness in decision-making in every level of the Academy. Late arrivals of the Soviet delegations to international conferences and delayed replies to the requests and proposals from foreign organizations occurred very often in those days.[32]

2.2 *The disparate guests*

In spite of such an inadequate arrangement, the Soviet Academy gathered a total of 41 foreign guests from 20 countries. Among those participants were Bertha Karlik (1904–1990. Austria), Erich Schmid (Austria), Tirane, Shqiperi (Albania), Kha Maung Maung (Myanmar, then Burma), Sandor Salay (Hungary), Istvan Covach (Hungary), Eberhard Leibnitz (1910–1986. German Democratic Republic; GDR), Heinz Barwich (1911–1966. First Director of Central Institute for Nuclear Physics of the GDR. Later immigrated to Western Germany in 1964), Wilhelm Macke (1920–1994. GDR), Jan Hendrik de Boor (Holland), Mahmud Ahmed (Egypt), Tagdich Shankar (India), Khanolkar Vasantra (India), Megnad Saha (1893–1956. India), Damodar Dharmananda Kosambi (1907–1966. India), Rudrendra Kumar Pal (1902–1991. India), Hisabi Mahmud (Iran), Benjamin Bloch (1900–1959. Israel), Siwabessy Gerrit (1914–1982. Indonesia), Arne Lundby (1923–1991; Norway), Thorbjorn Sikkeland (1923–2014. Norway), Teunis Johan Barendregt (1920–1991. Norway), Pawel Szulkin (1911–1987. Poland), Nilsl Doniel Fontel (1901–1980. Finland), Vaclav Votruba (1909–1990), Cestmir Simane (1919–2012), Jan Urbanez, Jaromir Maly (Czechoslovakia), Sigvard Eklund (1911–2000. Sweden. Later, Secretary-General of the International Agency of Atomic Energy from 1961–1981), Stig Melker Claesson (1917–1988. Sweden), Savic Pavle Savic (1909–1994. Then Yugoslavia), Drago Grdenic (1919–. Then, Yugoslavia), Yoshio Fujioka (Japan), Wang Ganchang (王淦昌. 1907–1998. China), Cheong Gun (丁根. North Korea), and others.[33] Some of them were already familiar to Soviet scientists; Erich Schmid had once worked together with several Soviet physicists and chemists in Germany. Pavle Savic stayed in the Soviet Union during World War II as a member of the Yugoslavian military mission. Heinz Barwich had been engaged in a Soviet research project related to nuclear development in Georgia for a long time and was awarded the Stalin Prize. Istvan Covach from Hungary had been captured and sent to a war prisoner camp in Krasnoyarsk for nine months from 1945 to 1946. Cheong Gun studied as a graduate student in Leningrad from 1948 to 1952 and passed his qualifying exams to become a candidate for a doctorate. Furthermore, some participants had pro-Soviet views; Bertha Karlik was an activist in an Austrian peace movement, while Covach from Hungary and Xue Yugu from China were communists.[34]

Of course, the organizers of the Session were the most pleased with Barendregt's (a scientist from Norway) admiration for the achievement of Soviet science. Judging from the achievements made by the United States in the field of atomic research that Barendregt was familiar with, he spoke more favorably of the Soviet achievements.[35] The organizers were, however, somewhat annoyed by some other guests who were difficult to deal with. Saha and Kosambi from India did not show much interest in the presentations at the Physics Section. Kosambi seemed not to be interested in the Session at all. At last, on July 3, he asked the organizers for relief from the duty to attend the Section on the occasion of Dr. Shankar's arrival at Moscow and also offered his own paper about feudalism in India and the ancient history of India.[36] His specialty was so far from physics. Savic and Grdenic also showed little interest in the presentations. They spent almost all day long at the entrance hall talking with Soviet scientists. On July 2, these Yugoslavian scientists dropped away from the Session, explaining that they would go to the Institute for Physics Problems and the Institute for Organic Chemistry.[37] Some delegates, including Lundby, Sikkeland, Barendregt from Norway, and Maung Maung Kha from Myanmar, enjoyed strolling in the town instead of attending the Session.[38] However, they alone should not be blamed. Simultaneous interpretation into English was available only in the Physics Section. The interpreters very often received a text just before a presentation, and they often had to interpret from page proofs. Some foreign participants who knew the Russian language repeatedly pointed out the errors in the simultaneous interpretation.[39] Fujioka later wrote that the simultaneous interpretation tended to "be interrupted" repeatedly.[40]

2.3 Excursion to Obninsk

The excursion to the Obninsk Atomic Power Station by bus took place on July 6, the day after the closing of the Session. During the observation of the atomic power plant, all the scientists who came from abroad showed great interest in the devices and works of the station. Wilhelm Macke from the German Democratic Republic eagerly wanted to take as many photographs of the nuclear power plant as he could, and ultimately took a total of 180–200 pictures. Yoshio Fujioka from Japan was also eager to take photographs and looked as if he were always waiting for a chance to enter the places where foreigners were not invited. Sigvard Eklund from Sweden annoyed the guides with a large number of detailed questions and eagerly sketched the location of the station, the main hall, the uranium blocks of the reactor, and the uranium rods. He tried to make his notes and drawings more accurate with assistance from Claesson and other delegates. Betha Karlik from Austria, who complained about the United States interfering with her trip to the Soviet Union, concentrated on taking notes in German of her observation of the station.[41]

Of course, almost all foreign guests admired the Soviet achievements in the field of atomic science and technology. It is, however, difficult to separate clearly sincere admiration from mere lip service. Even Fujioka, who repeatedly expressed admiration, refused a request of the Soviet journal, *Nauka i Zhizn'* (*Science and Life*), to write an article about his impression on the Session as well as a request for an interview from a correspondent of a news agency, TASS (*Telegrafnoe agenstvo*

Sovetsrogo Soyuza).[42] Western scientists were apparently cautious, fearing that they might be blamed by their governments and the public for possible involvement in pro-Soviet propaganda. Eklund seemed to be somewhat skeptical about the Soviet inclination toward giant projects like nuclear development at the cost of ordinary people's welfare. On the way back from the excursion, Eklund expressed his surprise at a stark contrast between the huge successes of the USSR in the field of nuclear construction and the impoverished views of villages, poor road conditions, and humble clothes of the inhabitants in the outskirts of Moscow.[43]

The Soviet interpreters, Petrovskaya and Racheev (their given names and patronymics are unknown), became aware of the suspicious behavior of some foreign guests. At midnight after returning from the atomic power plant, two Swedes and three Norwegians stayed together and compared their observations from visiting the station. They persuaded Fontell from Finland to join them. In the following day, he slept in the morning, saying that he could not sleep all night. Bloch from Israel was allegedly also involved in that work.[44] And then, Eklund hurried away from the Soviet Union to return to his own country on the morning of the 9th.[45]

Judging from the above-mentioned fact, the Foreign Affair Division of the USSR Academy of Sciences, which was in charge of the reception of foreign participants to the Session, considered it reasonable to assume a group of foreign scientists from the capitalist countries were collecting and processing all information that they managed to get through the Session and the excursion. Eklund's early departure could be explained well if it were supposed that he had been required to give the Western camp information about the institutes working in the field of nuclear physics that they observed in the Soviet Union. His rushed departure might be associated with the Geneva Conference. The Foreign Affair Division regarded Eklund as the central figure around whom Swedes, Norwegians, Finnish, and possibly other members of capitalist countries grouped.[46]

3 Soviet scientists meet big science: the First United Nations International Conference on the Peaceful Uses of Atomic Energy held in Geneva

The First United Nations International Conference on the Peaceful Uses of Atomic Energy was held in Geneva, Switzerland, from August 8–20, 1955, gathering approximately 1,400 representatives from 73 countries. In addition to these enrolled representatives, about 1,500 observers also took part in the Conference. More than 900 journalists were sent from a variety of countries to cover the Conference.[47]

Seventy out of 102 applications from the Soviet Union, including Ukraine and Belorussia, were accepted as the presentations to be read in the plenary meeting and various sections of the Conference. The number of their presentations exceeded those of the United Kingdom (65), France (61), and Canada (12).[48] A featured presentation by Dmitrii Blokhintsev (1907–1979), "The First Industrial Atomic Power Station of USSR and the Course of Development of Atomic Energy" (jointly with N.A. Nikolaev), attracted a great deal of interest from the audience.[49] A film describing the Obninsk Atomic Power Station received stormy applause.[50] The Presidium of the Soviet Academy of Sciences later proudly reported.

> The Soviet delegation at the International Conference was able to convincingly demonstrate the substantial progress and the achievements of Soviet science and technology in the field of the use of nuclear energy for peaceful purposes. The high scientific level of the reports of the USSR presented at the Conference, the readiness of Soviet scientists to share their experiences with scientists from other countries for the peaceful use of nuclear energy and the individual contacts which were established by the Soviet scientists with the foreign scientists at the Conference helped to raise the prestige of Soviet science and boast the authority of the Soviet Union.[51]

However, it must be taken into consideration that the Geneva Conference provided an arena for scientific intelligence gathering for both sides of the Cold War, as emphasized by John Krige,[52] and in this sense it was an extension of "a battlefield" in the Cold War. The Western camp headed by the United States took the initiative to organize the Conference. A French communist, Frédéric Joliot-Curie (1900–1958), was not allowed to participate in the Conference. No scientist was invited from the People's Republic of China or the German Democratic Republic, whereas some scientists from Taiwan (Republic of China) were allowed to participate.[53] The U.S. delegation, consisting of such eminent scientists as Walter Zinn (1906–2000), Isidor Isaac Rabi (1898–1988), Hans Bethe (1906–2005), Victor Weisskopf (1908–2002), and others, and who were accompanied by Chairman of the U.S. Atomic Energy Commission Lewis Strauss (1896–1974) and eight senators,[54] overwhelmed the Soviet delegation in several ways. They accomplished this with the amount of their presentations (170[55]); with the demonstration of driving a real experimental heterogeneous type 10,000-kilowatt class nuclear reactor[56] installed in the exhibition hall of the U.S. pavilion; and with their other impressive exhibitions. Ironically, in their demonstration of the nuclear reactor they showed the Cherenkov effect to the observers, which was discovered by the Soviet scientists.[57] Krige concludes, "The presentation of the U.S. reactor in Geneva was a masterpiece of marketing."[58] In conversation with some Soviet representatives, an American scientist expressed his indifference toward the construction of nuclear power stations due to the cheap and abundant coal resources in the U.S.,[59] which made a stark contrast with the Britons' enthusiasm for atomic energy expressed in John Cockcroft's speech at the conference, who was then the Director of the Atomic Energy Research Establishment.[60] Thus, the Soviet effort to build the first nuclear power station was somewhat sidestepped.

The Presidium of the Soviet Academy of Sciences could not help acknowledging their defeat, saying,

> The Conference also revealed that in a number of areas of peaceful uses of nuclear energy research and, especially, experimental works are conducted in

> the United States much wider than in the USSR. So was the development and creation of new types of experimental reactors, high-energy and high-voltage-current particle accelerators, new building materials and radiations in industry, agriculture, biology and medicine.[61]

Expressing a similar feeling, a physical chemist, Viktor Kondrat'ev (1902–1979), stated as follows: "Here it has been fairly told that in the United States the application of atomic energy for peaceful purposes is being developed more widely and the works are being done at a higher tempo."[62] A metallurgist, Aleksandr Samarin (1902–1970), softened Kondrat'ev's remark thusly,

> I would like to continue the remark of Academician V.N. Kondrat'ev in another aspect. . . . These [presentations read by the Soviet scientists at the Conference – *Ichikawa*] are the reports which were made by the 'landing' groups, i.e., very small groups of very qualified scientists. The scope of the development of the works on the peaceful use of nuclear energy is wider in our country, in my opinion, than in the United States.[63]

Notes

1 V.A. Sidorenko et al. (Ed), *K Istorii Ispol'zovaniya Atomnoi Energii v SSSR, 1944–1951* (*Dokumenty i Materialy*). Obninsk, GNTs Fiziko-energetichekii Institute, 1994, 126–127.

2 N.A. Dollezhal', *U Istokov Rukotvornogo Mira* (*Zapiski Konstruktora*) (*At the Sources of the Man-made World: Memoirs of a Designer*) (2-e izdanie). Moskva, Izdatel'stvo GUP NIKIET, 1999, 148–149, 217: Even in the very primitive stage, some of his colleagues like Savelii Feinberg (1910–1973) and Isai Gurevich (1912–1992) had already begun a trial for the development of a light-water reactor. The plan codenamed "Malyutka" was, however, long suspended due to passivity of the upper levels of government which wanted to concentrate all the potentials on making nuclear bombs (Pod red. Sidorenko i dr., *op. cit.*, in note 1. 107–113; A.P. Aleksandrov, *Akademik Anatolii Petrovich Aleksandrov: Pryamaya Rech'. – Academician Anatolii Petrovich Aleksandrov: Plain Speech.* Moskva, Nauka, 2001), 127–128.

3 Pod red. Sidorenko i dr., *op. cit.*, in note 1. 127: The codename "AM" is often referred to as an acronym standing for "Atom mirnyi (Atom peaceful)." Sonia Schmid insists that "AM" "originally referred to naval applications ("naval atom" or *atom morskoi*)" (Sonia D. Schmid, *Producing Power: The Pre-Chernobyl History of the Soviet Nuclear Industry.* Cambridge, MA, London, The MIT Press, 2015, 46). It seems, however, that each of such interpretations belongs to a sort of hindsight or a myth created *ex post facto.* As for the codenames of the early nuclear reactors designed by Nikolai Dollezhal' and his Institute, *Nauchno-Issledovatel'skii i Konstruktorskii Institut Energotekhniki* (NIKIET), see Pod red. V.K. Ulasevicha, *Sozdano pod Rukovodstvom N.A. Dollezhalya . . . : O Yadernykh Reaktorakh i Ikh Tvortsakh.* Moskva, Izd-vo GUP NIKIET, 1999.

4 V.N. Mikhailov i dr., *Atomnaya Otrasl' Rossii.* Moskva, IzdAT, 1998, 68.

5 Sidorenko et al. (Ed.), *op. cit.*, in note 1. 134–137.

6 Ibid., 140–142, 146, 147: Lev Kochetkov (1930–), a major specialist who had been in charge of operating that reactor for a long time, confirmed the fact of this conversion of the purpose of the "AM" development project in front of a camera

of NHK (Nihon Hôsô Kyôkai, a TV broadcasting agency of Japan) reporters. The interview was taken in September 18, 2014. The TV program including that scene was broadcasted on November 30, 2014, in Japan, under the title, "Kakuno Heiwa Riyô: Shirarezaru Mô-hitotsu no Tôzai Reisen (The Peaceful Use of Nuclear Power: Another Unknown Cold War)."

7 Cited in V. Vovulenko, "Vstpitel'naya stat'ya (Introductory Article)"//Dzh. Allen, *Atomnaya Energiya i Obshchestvo* (A Russian version of James Allen, *Atomic Energy and Society*. New York, International Publishers, 1949.) Moskva, 1950, 5–19.

8 Kakuzô Maeshiba, *Genshiryoku to Kokusaiseiji: Kyôzon ka Kyômetsu ka* (*Atomic Energy and the International Politics: Coexistence or Co-extinction?*). Tokyo, Tôyô-Keizai-Shipô sha, 1956, 69.

9 G.M. Malen'kov, "Ochyotnyi doklad Tsentral'nogo komiteta VKP(b) XIX s''ezdu partii." http://stalinism.ru/dokumentyi/materialy-xix-s-ezda-vkp-b-kpss.html?tmpl=component&print =1&page=.

10 Sidorenko et al. (Ed.), *op. cit.*, in note 1. 13–14.

11 Ibid.

12 L.D. Ryabev (Ed.), *Atomnyi Proekt SSSR: Dokumenty i Materialy*, T. II. Kn. 2. Moscow, Sarov, Nauka-VNIIEF, 2000, 428–436.

13 Ibid., 434.

14 Ibid., 434.

15 As for the unique history of Russian/Soviet Academy of Sciences, see Hiroshi Ichikawa (Ed.), *Kagaku no Sanbô-honbu: Rosia/Sorenpô Kagaku Akademî ni kansuru Kokusai Kyôdô Kenkyû* (*The General Staff of Science: An International Joint Study on Russian/Soviet Academy of Sciences.*) Sapporo, Japan, Hokkaido University Press, 2016.

16 Rossiiskii gosudarstvennyi arkhiv sotsial'no-politicheskoi istorii (RGASPI), Fond (F.) 17, Opis' (Op.) 133, Delo (D.) 171, 2–3: Sonja Schmid regards that the political behaviors taken by the then leading scientists were in many cases oriented to the rehabilitation of fundamental sciences [Sonja D. Schmid, "Defending (Scientific) Direction: Soviet Nuclear Physics and Reactor Engineering during the Cold War." in Naomi Oreskes and John Krige (Ed.), *Science and Technology in the Global Cold War*. Cambridge, MA, London, The MIT Press, 2014, 317–342]. However, the unique organizational characteristic of the development of science in Imperial Russia and the Soviet Union must be also taken in consideration, in which the Academy of Sciences remained as a powerful, practical organization for a long time contrary to the Western experiences, as I argued in "Introduction" to this book. Some kind of arguments presuming some conflicts between the fundamental research and the applied research must be reconsidered because of the ambiguity of the boundary and because of the mutual dependence between them. The latter point was recently highlighted by Hikari Mizusawa on the instance of Japanese wartime research [Hikari Mizusawa, "Nicchû Sensô-ka niokeru Kiso Kenkyû Shifuto Kagaku Kenkyû-hi Kôfukin no Sôsetsu (Basic Scientific Research during the Second Sino-Japanese War: The Establishment of the Subsidiary Fund for Scientific Research)." Japan History of Science Society. *Kagakushi Kenkyu* (*Study on the History of Science*), Vol. II-51, No. 264, 2012, 210–219].

17 Alexei B. Kojevnikov, *Stalin's Great Science: The Time and Adventures of Soviet Physics*. London, Imperial College Press, 2004, 123–146.

18 Ryabev (Ed.), *op. cit.*, in note 12. T. II, Kn. 3, 2002, 93–97.

19 A.M. Petros'yants, "Desyatiletie yadernoi energetiki." *Atomnaya Energiya*, Vol. 16, No. 6, 1964, 480.

20 D.I. Blokhintsev, N.A. Dollezhal' and A.K. Krasin, "Reaktor atomnoi elektrostantsii AN SSSR." *Atomnaya Energiya*, Vol. 1, No. 1, 1956, 10–14.

21 G.N. Ushakov i dr., "Opyt ekspluatatsii Pervoi v mire atomnoi elektrostantsii AN SSSR." *Atomnaya Energiya*, Vol. 16, No. 6, 1964, 485–488.

22 As for the interview, see the note 6.
23 State Scientific Centre of the Russian Federation – Institute for Physics and Power Engineering named after A.I. Leypunsky, *The First in the World Nuclear Power Plant (To the 60-th Anniversary of Commissioning)*. Obninsk, 2014, 190–191: The United States dropped the atomic bombs not on Germany but on Japan and repeated hydrogen bomb tests in the Pacific Ocean region. In short they had never used nuclear weapons toward any white (Caucasoid) people. That fact was a threat for some of the leaders of Asian countries, like Nehru, who felt the racist trend in the US nuclear strategy. They became deeply interested in the problems related to nuclear science and its development in the countries other than the United States (See, Matthew Jones, *After Hiroshima: The United States, Race and Nuclear Weapons in Asia, 1945–1965*. Cambridge, Cambridge University Press, 2010).
24 Arkhiv Rossiiskoi Akademii nauk (ARAN), Fond (F.) 579, Opis' (Op.) 18 (1955), No. 6, 4: As for rigid confidentiality of scientific research which obstructed international exchanges of scientific knowledge for a long time in the Soviet Union, a commission was formed with a physicist, Nikolai Dobrotin (1908–2002), in chair in the meeting of the Presidium of the USSR Academy of Sciences on May 28, 1954, to make a new proposal to the Academy on the criteria of confidentiality in scientific research. Eventually, on September 2, 1955, the USSR Council of Ministers adopted a proclamation for the substantial relaxation of foreign travels and other overseas exchanges of scientific information by scientists and other specialists (ARAN, F. 2, Op. 6, D. 170, 197).
25 A.N. Nesmeyanov, "Introductory remarks to the Session of the Academy of Sciences of USSR on the Peaceful Use of Atomic Energy on July 1st 1955." in Akademiya Nauk SSSR (Ed.), *Sessiya po Mirnomu Ispol'zovaniyu Atomnoi Energii* (The Academy of Sciences of the USSR (Ed.), *The Session on the Peaceful Use of Atomic Energy*), July 1–5, 1955, Vol. 1. Moscow, Izd-vo AN SSSR, 1955, 9.
26 ARAN, F. 579, Op. 1, No. 19, 153–162.
27 ARAN, F. 579, Op. 1, No. 20, 4.
28 Ibid., 122.
29 Ibid., 112.
30 Ibid., 153.
31 Fujioka, Yoshio, "Suisen no Ji (Recommendatory Address)," in Soren Kagaku Akademî (Ed.), *Genshiryoku Heiwa Riyô Kaigi Hôkoku Ronbun Shû (Kôgaku Bukai)*. Tokyo, Maruzen, 1956, No page number.
32 ARAN, F. 579, Op. 18, No. 6, 4–9.
33 ARAN, F. 579, Op. 1, No. 19, 44, 59, 145–150.
34 Ibid., 36, 38, 40–41, 44 and 59.
35 Ibid., 38.
36 Ibid., 53.
37 Ibid., 54.
38 Ibid., 56.
39 Ibid., 61.
40 Fujioka, *op. cit.*, in note 31.
41 ARAN, F. 579, Op. 1, No. 19, 71, 74–75: It seems that the participants were shown a film describing the operation of the Obninsk Atomic Power Station during the Session before the excursion (Ibid., 52).
42 Ibid., 52–53.
43 Ibid., 70.
44 Ibid., 73: The importance of "an awareness of foreign scientific developments" was emphasized in the Western camp from the early days of the Cold War. For example, Krige pointed out the fact that "the use of international scientific exchange as an instrument of scientific intelligence gathering was officially promoted and sanctioned in the classified appendix to a report prepared by a

panel established by Lloyd Berkner [1905–1967; a geophysicist – *Ichikawa*] at the request of the State Department. Titled *Science and Foreign Relations*, and partially released in May 1950, the Berkner report insisted that an awareness of foreign scientific developments was crucial to the progress of American science" (John Krige, "Atoms for Peace, Scientific Internationalism, and Scientific Intelligence," *Osiris*, Vol. 21, No. 1, 2006, 166).

45 ARAN, F. 579, Op. 1, No. 19, 76.
46 Ibid., 74–76.
47 ARAN, F. 694, Op. 1, D. 101, 3.
48 ARAN, F. 2, Op. 6, D. 201, 9, 98.
49 ARAN, F. 694, Op. 1, D. 101, 7–8.
50 ARAN, F. 2, Op. 6, D. 201, 106.
51 Ibid., 9–10.
52 Krige, *op. cit.*, in note 44. 165–168.
53 ARAN, F. 694, Op. 1, D. 101, 3–4: The hostile atmosphere compelled the United States to withdraw the reports sent by the Taiwanese scientists from the proceedings (ibid., 3–4).
54 Ibid., 96.
55 Ibid., 98.
56 Ibid., 11.
57 ARAN, F. 2, Op. 6, D. 201, 105.
58 Krige, *op. cit.*, in note 44. 175.
59 ARAN, F. 2, Op. 6, D. 201, 117.
60 Ibid., 100.
61 Ibid., 10.
62 Ibid., 138.
63 Ibid., 139–140.

7 "A double-edged sword"

Radiation studies and the normalization of biological sciences

In the second half of the 1950s, Soviet scientists stood at the crossroads in understanding radiation effects on the living body: Their perspective was changing from that of a mere follower of Western studies to that of a radical critic. Looking for a disregarded context in which research on radiation effects emerged as a major topic in Soviet science, this chapter explains the indigenous needs for such research in the Soviet Union as well as its early development by shedding light on radiation casualties at a nuclear development center, Chelyabinsk-40 (later, -65) and hazard studies conducted there in the early years. After that, we trace the path along which many physicists and other scientists came to take actions for the normalization of biological sciences.

1 The Soviet nuclear weapons development and radiation exposure

To develop the atomic bombs as soon as possible, the Soviet government hastily built two huge industrial complexes, one for producing plutonium and the other for uranium enrichment, as well as one large scientific research center, a vast test site and other facilities in a very short time. In those early days, administrators, scientists, and engineers involved in that project, working under enormous pressure to achieve maximum success in the shortest possible period, made various errors and oversights due to haste. One of the most serious consequences was careless handling of radioactive materials.

1.1 Radiation exposure in Chelyabinsk-40 in its early years

Angelina Gus'kova (1924–2015), a specialist of radiation medicine, has offered rare insights in radiation hazards at Industrial Complex No. 817 (codenamed as Chelyabinsk-40, later renumbered -65) built for plutonium production. She read a paper prepared jointly with Grigorii Baisogolov (1921–2003) on the problem of cure for heavy radiation sickness at the International Conference on Peaceful Uses of Atomic Energy held in Geneva in 1955.[1] According her witness, during the first decade of its operation, from 1948 to 1958, a total of 42 persons suffered from serious radiation sickness. Seven of them died. Two-thirds of the exposure incidents happened at Works No. 25 in Plant B where plutonium was separated; one-fifth at Works No. 20; one-tenth at Plant A with the graphite-moderator type nuclear reactor; and one-twentieth at various construction sites.[2]

The most serious incident occurred on June 17, 1948. Reactor A at Industrial Complex No. 817, the first Soviet full-scale reactor for plutonium production, was fully loaded with fuel blocks in every operating channel and reached the critical state three days later. However, after a brief moment of joy, the reactor broke down on the next day. Radioactive products equivalent to approximately 300 doses (one dose: 1 coulomb /1 kg) were released into the air, while a hole on the graphite base was clogged up with slag. The heat expansion of graphite and uranium had prevented the smooth cooling of water supply. Igor' Kurchatov (1903–1960), leader of the Soviet nuclear weapons development, nevertheless ordered its continuous operation as "an exceptionally dangerous, but only one feasible solution," without fundamental repair or cooling of the clogged hole. Immediately after the reactor began to operate again on July 12, however, slag built up in another hole. In order to remove it, many workers had to work under dangerous conditions with radioactive particles: I.P. Frolov, for example, was exposed to radiation up to 26 roentgen; B.S. Petrov, 46 roentgen; M.N. Pichugin, 53 roentgen; and L.I. Zakharov, 108 roentgen.[3]

It was not until the end of 1948 that Kurchatov and other leaders of the Soviet nuclear weapons development decided to fix the problem. On January 20, 1949, the operation of the reactor was suspended, with a total of 39,000 blocks of uranium fuel unloaded. It took almost two months before the reactor resumed its operation on March 26. Ironically, after the repair, the operation of the reactor had to be limited. This was because operations had to be kept to lower levels to keep the temperature of graphite below 330 centigrade in order to avoid expansion from heat. During the repair work, workers involved in the reactor operation were exposed to radiation at a high average annual rate of 93.6 rem (936 mSv) in 1949.[4]

At Plant B, where a large amount of radioactive materials were treated in the process of plutonium separation for fission bombs, workers also risked their health due to serious radiation hazards. Levels of exposure were dangerously high especially when much of the reactor's equipment was upgraded for better efficiency. This also happened when a great deal of the fissile materials were replaced due to a series of erosion events in 1950 and 1951. Eventually, a scientist working at the plant named Boris Nikitin died from acute radiation sickness in 1952 at the age of 46. Four years later, in 1956, Aleksandr Ratner, who actually played a leading role in the plant, also fell victim at the age of 50. The total number of persons suffering radiation injuries at work during the first two decades since the opening of the plant amounted to 2,089 of approximately 11,000 persons working at this plant then.[5]

The most serious incident at Plant B happened on March 15, 1953. A crane broke down, and highly concentrated plutonium solution fell out of a container. This nearly reached criticality. Two workers in charge of the disposal of radioactive wastes, whose family names were Ershov and Timazeev, died from serious radiation exposure.[6] Even when no accident happened, exposure levels remained alarmingly high. The annual average doses received by a worker of Plant B reached 48.0 rem in 1949, 94.0 rem in 1950, and 113.3 rem in 1951. Radiation hazards even spread outside the nuclear complex. Since the very beginning of its operation, low-level radioactive wastes, which were somewhat diluted but untreated, were discharged into the Techa River.[7]

Many scientists and administrative staff involved in the construction and operation of Industrial Complex No. 817, including Kurchatov, Director of the complex Efim Slavskii (1898–1991), and others, were also heavily exposed to radiation.[8] To facilitate the investigation and treatment of those exposed to radiation as well as the development of protection measures, a field hospital opened in 1949, followed by the establishment of a field branch of the Institute of Biophysics of the USSR Ministry of Health (and the Soviet Academy of Medical Sciences) in 1952.[9] The medical doctors, researchers, and medical workers there, however, were ordered by the authority to keep absolutely silent about their secret tasks. It was strictly forbidden for them even to record the facts and the numerical data concerning radiation sickness. Therefore, they had to use the other words indicating radiation sickness, like "myasthenia-vegetative syndrome" and others.[10]

1.2 The East Ural Radioactive Trace

On September 29, 1957, at 4 pm, problems arose with the cooling system of a concrete container, with the capacity of 300 m^2 holding 70–80 tons of the nitric acid acetate solution of high-level radioactive waste. The solution overheated and exploded. Approximately 10% of the contained radioactive material (^{90}Sr, ^{90}Y, ^{95}Zr, ^{95}Nb, ^{144}Ce, ^{144}Pr, ^{106}Ru, ^{106}Rh, ^{137}Cs, etc.), equivalent to 20 million curies, was released into the air and carried northeast by the wind. The Central Laboratory of Industrial Complex No. 817 was ordered to take primary responsibility for investigating casualties. Also mobilized were the medical corps of the complex: the field branch of the Institute of Biophysics, the branch of the Institute of Radiation Hygiene of the USSR Ministry of Health, the Institute of Applied Geophysics, and the Institute of Geochemistry of the Soviet Academy of Sciences.[11] Estimates later indicated that, in total, 6,600–8,300 people were likely to have been exposed to radiation. At first, the USSR Ministry of Medium-size Machine-building and the Ministry of Health answered an inquiry made by the Chairman of the Sverdrovsk Oblast Executive Committee with reassurance, stating: "There is no sign of any serious pollution in the territory of the Oblast." Such optimistic and irresponsible attitudes of the government, however, led to a serious delay in taking countermeasures. Only in April 1958, almost six months after the accident, was the first such measure – numerical control on the number of livestock and that of diseases among the inhabitants – finally taken in the Kamensk region. Belatedly, further actions followed: the disposal of the contaminated crops; the prohibition of consuming food raised in the polluted area; and the evacuation of the inhabitants from the highly radioactive zone. External exposure doses received by the residents had been continuously measured until 1963. What is interesting is that those who were engaged in the measurement of exposure doses actually invented their own method of gauging exposure. This indicates the seriousness of the internal radiation exposure caused by the accident. First, the magnitude of the transfer of radionuclides from soil to crops was estimated. Then the average daily diet for every age group of the inhabitants was calculated. With these calculations in hand, experts made estimates about the annual average intake of radionuclides into the human body.[12]

2 Indigenous development of research on biological effects of radiation

2.1 Valentin Khokhryakov and his team

Here we will trace the progress in the indigenous and original studies on radiation effect made in the Soviet Union until the summer of 1958.

For many years in his professional career, Valentin Khokhryakov (1928–), a scientist in the field branch of the Institute of Biophysics at Cheryabinsk-40 (later, -65), supervised the development of methods to measure radiation effects on human bodies as well as that of means of protection from exposure.[13] At first, his team concentrated their efforts on measuring external exposure to β- and γ-rays and estimating their effects. As early as in the 1950s, however, they discovered surprisingly large amounts of plutonium from inside the bodies of dead workers at the complex.[14] Specifically, they used the α-ray spectrometer to detect this type of radiation emitted by americium-241 and some plutonium isotopes in the ashes of bones of the dead. Such measurements were conducted twice in the decade from 1949 to 1959.[15]

2.2 "Session for the Peaceful Uses of the Atomic Energy" in 1955

In August 1955, the International Conference on Peaceful Uses of Atomic Energy was held in in Geneva. A month earlier, July 5–7, the Soviet Academy of Sciences hosted a large domestic conference as a preparatory event to that in Geneva. What is noteworthy here is that one of the sections was dedicated to biological sciences. In his presentation, the section's organizer Leon Orbeli (1882–1958), a leading physiologist, emphasized the importance of research on biological effects of radiation in the nuclear era, saying: "No matter whether research and utilization of atomic energy is carried out for peaceful or other purposes, this energy has some effects on the human being and any living body."[16] Of particular importance were a number of new findings reported during the Session. Pavel Minaev found that radiation exposure triggered a shift in the chicken's nerve center to some excitement and some morbid fluctuations.[17] Orbeli also refers to a similar research result treating with ovulation of frogs irradiated with gamma-ray.[18] A.G. Pasynskii made another report, stating: "When radioactive ray goes through the water, a large amount of free radicals (especially, OH radicals and O_2H radicals) are generated. And, some chemical reactions, especially some oxidizations are brought about by means of the interaction of those radicals and the matters in the solution."[19] Orbeli and Aleksandr Kuzhin (1906–1999) also confirmed his findings.[20]

2.3 Medical scientists and others

In the following year, from January 30 to February 4, 1956, the USSR Ministry of Health held the First All-Union Conference on Medical Radiology. A total of 232 papers were presented at the meeting, which a review article described as providing "the research results made by the Soviet scientists concerning effects of ionizing radiations onto organism, the uses of atomic energy to biology and medicine, radiation measurements and hygienic problems of radiology."[21] Mikhail Livanov (1907–1986) discussed in detail radiation effects on nerve activities, a topic which Western scientists largely overlooked in those days.[22] A.A. Bagdasarov and L.S. Rogachev reported on their practices in the medical treatment of patients suffering acute and chronic radiation sickness.[23] An authoritative medical doctor, Avgust Letaved (1893–1984) also made an urgent appeal to advance basic research as an aid for establishing standards on permissible doses.[24]

Soviet research made further progress by the time when another major conference – the All-Union Scientific and Technical Conference on the Applications of Radioactive and Non-Radioactive Isotopes in National Economy and Scientific Research – was held on April 4–12, 1957. During this week-long conference, attended by more than 3,000 participants representing 1,016 organizations, a total of 444 scientific reports were presented. One of the major topics was the problem of what was later called global fallout, the radioactive debris scattered worldwide due to atmospheric nuclear testing.[25] V.P. Shvedov and L.I. Gedeonov reported on the pollution of the biosphere around Leningrad due to this globally circulating contaminant.[26] V.M. Klechkovskii and I.V. Guryakin reported on the contamination of soil with radioactive fallout and the absorption of radioactive strontium by different parts of plants.[27] In addition, G.G. Tinyakov and M.A. Arsen'eva presented the result of their research on the cytogenetic consequences of radiation effects as observed in the spermatozoa formation of the apes.[28] Most of these presenters later contributed to the book published in 1959, A.V. Lebedinskii (Ed.), *Sovetskie Uchyonye ob Opasnosti Ispytanii Yadernogo Oruzhiya* (Moscow, Atomizdat. 1959; *Soviet Scientists on the Danger of Nuclear Weapon Tests*).

What is interesting about Soviet radiation studies in those days is that the dangers of radiation sicknesses were openly talked about in public. A famous hygienist, Mikhail Pobedinskii published a popular reference book entitled *Radiation Sicknesses* in 1957.[29] This book was among the first that introduced the Soviet people to the tragedies of Hiroshima and Nagasaki.

3 The Academy of Sciences: its late start and its trial to catch up

3.1 The backwardness of the Soviet study on radiation effects

The Soviet Union succeeded in a series of nuclear weapon developments in a very short time since the success in the explosion test of the first Soviet atomic bomb,

RDS-1 on August 29, 1949, to the success of the first thermonuclear bomb, RDS-6 on August 12, 1953. Whereas the Soviet Union was engaged in producing and storing many nuclear weapons in reaction to the U.S. nuclear weapons programs, on the other hand, they conducted a campaign for "the peaceful use of atomic energy," with an anti-American propaganda, so as to appeal to domestic and overseas public sentiment.[30] Along such a policy line, the Soviet Academy of Sciences held a large-scale scientific session, Session on the Peaceful Use of Atomic Energy on July 1–5, 1955, the achievements of which were published in five-volume proceedings.[31]

What is noteworthy here is, as mentioned before, that one of the Sections was dedicated to biological sciences and a number of new findings were reported during the Session.[32]

Subsequently to this Session, the First United Nations International Conference on the Peaceful Uses of Atomic Energy was held in Geneva August 8–20, 1955. The Soviet government dispatched a large-scale delegation to that Conference. Obtaining some admiration of the participants for their success with the world's first nuclear power plant at Obninsk near to Moscow which began operations June 27 of the previous year, they, however, became aware of their backwardness in this field. The Conference made them realize that their achievements were "made by the 'landing' groups, i.e., very small groups of very qualified scientists (the remark in a meeting of the Presidium of the Soviet Academy of sciences by a metallurgist, Aleksandr Samarin – 1902–1970 –)" and their research fell far behind the "big science" in the United States and the United Kingdom in its scale.[33] June 15 and 22 of the following year, the Presidium deliberated on and adopted a large, comprehensive plan of its Department of Physical and Mathematical Sciences in the field of nuclear research for the following five years and, as part of a series of the measures for that, decided to establish a laboratory for radiation genetics in the Institute of Biophysics.[34]

The Bikini incident made the effects of radiation on living bodies a central issue of international politics.[35] Nevertheless, the Soviet Union was late and unprepared in this field. In the first place, the Soviet Union lacked the material conditions for utilizing radioactive isotopes for such research. When the Presidium of the Academy held a meeting on November 29, 1957, Chief Scientific Secretary of the Presidium of the Academy of Sciences, Aleksandr Topchiev (1907–1962), read his report on his attendance at the UNESCO (the United Nations Educational, Scientific and Cultural Organization)'s International Conference on Radio-isotopes in Scientific Research held in Paris on September 9–20. In it, he complained:

> Radiometric apparatuses, dosimeters, and other electronic physical apparatuses made in our homeland do not meet the necessary level in their variety, quality and, very often, technical performance for scientific research treating with isotopes and nuclear radiation. Due to the lack of protective equipment and clothing for the workers working with isotopes and radiation, many

> laboratories of various ministries and agencies are currently closed by the orders of the State Sanitary Inspectorate.[36]

Furthermore, at a Presidium meeting on December 6 in which a plan strengthening research activities for the solution of "structural, physical-chemical bases of fundamental biological phenomena" was adopted, Gleb Frank (1904–1976), a leading biophysicist of the time, admitted these dismal situations in his address, pointing out the shortage of the research activities related to electronic microscopes in the Soviet Union and stating, "We are left so far behind in the broad front of research for the analysis of the molecular structures of the living tissue."[37]

3.2 *The revolt of five prominent physicists*

In the Autumn of 1956 elections were held for the post of President of the USSR Academy of Sciences. A prominent chemist and the incumbent President Aleksandr Nesmeyanov (1899–1980, President of the Academy in 1951–1961) was the only candidate and was elected to the presidency for one more term. During the process of the election, in order to nominate a presidential candidate a general meeting of each Department of the Academy of Sciences was convened. At a Presidium meeting of the Academy on October 12, Topchiev reported about the happening at the general meeting of Department of Physical and Mathematical Sciences: The five leading physicists, Igor' Tamm (1895–1971. A Nobel laureate in Physics in 1958), Mikahil Leontovich (1903–1981), Lev Artsimovich (1909–1973), Grigorii Landsberg (1890–1957), and Pyotr Kapitsa (1894–1984. A Nobel laureate in Physics in 1978) proposed to postpone presidential elections until the Annual meeting of the Academy of Sciences in February of the following year, because a new research institute for genetics with a notable anti-Lysenko biologist, Nikolai Dubinin (1907–1998), as its Director was not established yet and Nesmeyanov didn't have a plan for radical change in biological sciences.[38] By doing so, they demanded Nesmeyanov to make clear the programmatic course for the improvement of the situation. Although their proposal was rejected in a Presidium meeting, this episode shows the frustration of the leading physicists with the very slow progress of building up the foundation for the radiobiological research in spite of the intensification of the nuclear armament race and the rapid growth of molecular biology at the time.

3.3 *Establishment of the Institute of Radiation Biology and Physicochemical Biology*

Such an objection by prominent physicists might have made some difference. On March 29, 1957, with the aim of making a breakthrough in the fields of radiation biology, biophysics, and chemical and physical research of isotopes, the Presidium of the Soviet Academy of Sciences made a decision to establish a new, large-scale, and integrated research center for studies in those three fields.[39] In discussions for this resolution, Kapitsa enthusiastically appealed to his colleagues in the Presidium, so much so that he declared:[40]

> I think that I must bluntly say, entirely unashamedly, that the future war would be decided by biologists, by no one else, if it broke out indeed or not! No physicists can now decide, but biologists will decide what will be the consequence of a nuclear war! We have been timidly left behind here. Yes, we were timid! We must not consider this problem in the same way with others. We need this type of institute for genetics, an institute for radiation genetics. . . . We must establish this kind of institute no matter what means we use. We must not dilly-dally like by now.

On April 26, less than a month after the Presidium meeting, the Institute of Radiation and Physicochemical Biology of the Academy of Sciences of USSR was founded with a famous biochemist, Vladimir Engel'gardt (1894–1984), as Acting Director. The Main Directorate of Atomic Industry (perhaps, correctly – the First Main Directorate affiliated to the Council of Ministers of USSR) also took part in its foundation, taking charge of the site selection for the institute and also of designing the buildings together with the Academy of Sciences.[41]

The decision to establish this institute, however, was not implemented for more than a year due to the lack of some necessary prerequisites. Engel'gardt gave a speech at the Presidium meeting on May 16, 1958, to discuss causes of the delay, saying,[42]

> Abnormal situation with working spaces in the institutions of the Department [of Biological Sciences. . . *Ichikawa*] prevents them from strengthening research activities in the fields left behind in biology and reduces the productivity of institutions. Especially intolerable is the situation with working spaces in the Laboratory for Radiation Genetics and Laboratory for Helminthology. Due to lack of spaces specialized for the use of isotopes and radiation in the institutions of the Department for Biological Sciences the research related with them is not carried out in proper conditions, which repeatedly led to their prohibition by District Sanitary Inspectorate.

Besides this, in the same meeting, V. S. Rusinova pointed out difficulty in decision-making in the Department for Biological Sciences. She complained;[43]

> Shouldering too heavy burden of obligations onto the members of the Bureau of the Department reflects in the activities of the Department. Usually at a Bureau meeting only half of its members (5–6 people) or sometimes a smaller number are present. . . . The today's defect of the scientific and organizational activities of the Department is the lack of major discussions on various important biological problems.

Vice president of the Academy, an economist Konstantin Ostrovityanov (1892–1969), however, reminded the Presidium members of "a kind of opposition, as seen symbolically in the various fields of biology, against the new direction in which a wide range of physical and chemical methods must be adopted."[44] It was needless to say that the "opposition" meant maneuvers staged by Lysenkoites (this aspect will be examined later in this chapter).

On January 31, 1958, "the Commission on Radiation Biology" affiliated to the Presidium was set up. It consisted of 25 members, including the nuclear physicist Anatolii Aleksandrov (1903–1994. Later, President of the Academy for 1975–1986), the leader of Soviet nuclear development project, Igor' Kurchatov, Orberi, Nikolai Semyonov (1896–1986. A Nobel laureate in Chemistry in 1956), Tamm, Engel'gardt, Dubinin, Gleb Frank, and other eminent scientists. The commission was delegated authority to coordinate all the activities related to "the effects of nuclear radiation to living bodies and heredity" and those carried out in the Academy of Sciences.[45]

For the radical promotion of the research activities in two areas of "practical, vital importance" as identified by the institute (i.e., the problems of radiation damage, their elimination and prevention, and the problem of malignancy or, in other words, malignant growth of tumors), Engel'gardt and other organizers of this institute made great efforts in recruiting several able specialists necessary for their purposes. Some of these specialists include Aleksandr Braunshtein (1902–1986. A full member of the USSR Academy of Medical Sciences, and a foreign member of the National Academy of Sciences, USA, in 1974), Professor V.I. Tovarnitskii, Yakov Varshavskii (1917–?. From the Physicochemical Institute named after L.Ya. Karpov. Later, in 1991, he moved to Salt Lake City), Maksim Meisel' (1901–1987, a microbiologist-cytologist), Lev Tumerman (1907–1986. Later, a coauthor of the comprehensive voluminous work, *Molecular Biology*. In 1974, he emigrated to Israel), Ivan Utkin (1921–?. Then Director of Sukhumi Experimental Pathology and Therapeutics Institute), and others. They received their first salaries from the institute in the last week of May 1959.[46] The organizers of the institute also put great efforts in the preparation of research equipment. Thanks to their efforts, the institute was equipped with one Japan-made scanning electron microscope, 17 centrifuges, one Britain-made refrigerated centrifuge, one Japan-made registering spectrophotometer, 21 microscopes, three fraction collectors, three spectrophotometers, some micro-photometers, universal photometers, micro-x-ray gauges of the conversion type, and other devices in 1959. In addition, they also succeeded in securing 47 rooms, of which 26 rooms were well equipped, in a complex of buildings, where the Institute of Mining, the Institute of Biochemistry, and others were also located, as some parts of the Institute of Mining moved to another space.[47]

3.4 *The visits of the Western scientists to the Institute of Radiation Biology and Physicochemical Biology in 1960 and their impacts*

Soon after the actual initiation of the research activities of the Institute of Radiation Biology and Physicochemical Biology, the Academy of Sciences of the USSR organized a large-scale international symposium titled "The International

Symposium on Primary and Initial Effects of Ionizing Radiations on Living Cells," obtaining the auspice of UNESCO and co-sponsorship of the International Atomic Energy Agency (IAEA). An eminent Belgian scientist, Z.M. Bacq, who was a foreign member of the USSR Academy of Sciences, took the post of President of this symposium held in Moscow on October 18–22, 1960. Seventeen Western scientists like P. Alexander, Tor Brustad (1926–2016), Louis Harold Gray (1905–1965), Alexander Hollaender (1898–1986), Ernest Pollard (1906–1997), Cornelius Tobias (1918–2000), M. Errera, and others, six Czechoslovakian scientists, and 21 Soviet scientists such as Gleb Frank, Andrei Lebedinskii (1902–1965), Meisel,' A.G. Pasynskii, and others took part in this symposium. A total of 24 reports were read there: seven from the Soviet Union, two from Czechoslovakia, five from the United States, three from Great Britain, one from Belgium, France, Netherlands each, and one joint report of Britain and Belgium.[48]

The exciting presentations were followed by a heated general discussion. Along with Pasynskii, Lev Blyumenfel't (1921–2002), a professor at the Moscow State University, criticized the Western method of detecting free radicals by using an electron spin resonance spectroscope (the ESR method), saying, "It is, of course, possible that many injuries to biological macro-molecules caused by ionizing radiation are not connected with the appearance of unpaired electrons. In this case, the ESR method can be of no assistance."[49] Many Western scientists objected to them one after another. For example, Alexander said,

> But I would like to point out that our data do invalidate the argument used by Gordy and Shields to establish that the electrons wander to the sulphur in proteins. Their argument rested entirely on the fact that the ESR of irradiated protein and of irradiated cysteine were similar. We have now shown this similarity to be an artefact. Our chemical data indicate that there is no preferential attack on cysteine and it is now up to provide the evidence.[50]

Dubinin enumerated his doubtful points on the Western argument concerning "self-repair of genes" after the radiation exposure and requested the participants to reconfirm the following three principles: "1. The absence of a threshold radiation dose. 2. The existence of a direct dependence of the frequency of the mutations upon the dose, and 3. The cumulative effect of small radiation doses."[51] However, in opposition to him, Bacq, the President of the Symposium, replied, "Chromosome breaks are not only the result of direct radiation effects. They are the result also of secondary effects and can be modified by metabolic activity of the cell."[52] At the end of this exchange of words, Blyumenfel't and Dubinin became silent in front of a surprisingly large amount of data on which the Western arguments were based. In spite of the strenuous efforts by the Soviet scientists,[53] the Western scientists took the leadership in the symposium.

The Institute of Radiation Biology and Physicochemical Biology was involved in the organization and conduct of that symposium. It hosted several foreign guests from the Western countries in 1960; some of them were Hollaender,

Cornelius Tobias (also participated in the symposium), Paul Doty (1920–2011), Stanley Auerbach (1922–2004), Herman Kalcker (1908–1991), Norbert Wiener (1894–1964, famous as the founder of "Cybernetics"), Jacques Monod (1910–1976. A Nobel Prize laureate in Physiology or Medicine in 1965), and others.[54] Almost all of them gave their lectures at the institute. The lectures given by Doty (on the physicochemical research of nucleic acid), Tobias (on the effects of radiation to yeast cells), Auerbach (on inducers of mutation), and Keith Porter (on microscopy) highly impressed the researchers of the institute.[55]

However, among the foreign guests to the institute, Hollaender brought about the most vivid impact on the Soviet study of radiation effects. The annual report of the institute in 1960 says,

> Very interesting and instructive was the report of Hollaender (USA) on the organization of research in the Oak Ridge Laboratory. Prof. Hollaender stressed that despite the enormous importance of the research work on the effects of radiation on biological objects, about a half of the research activities is occupied by work related to the study of axial processes occurring under normal conditions of life of organisms in the laboratory (which is one of the major scientific centers of the United States).[56]

Hollaender was well known as the leading radiation biologist of the United States in those days. He worked in the Oak Ridge National Laboratory as the Director of its Biology Division for a couple of decades. He once visited the Soviet Union to study the mitogenic radiation hypothesis proposed by a Russian biologist, Aleksandr Gurvich (1874–1954), in 1931. After his return to the USA, Hollaender engaged himself in the research of the effects of radiation on genes, especially the damages and repairs of DNA. In the Oak Ridge National Laboratory, he made great efforts to obtain a large amount of research funds and to organize large-scale experimental research projects.[57] As the Director of the Biology Division at Oak Ridge, he supported and promoted the so-called Megamouse Experiment, a magnificent experiment on the mutation frequencies of mice caused by radiation in low doses, conducted by W.L. Russell and his team using 7 million mice overflowing the vast experimental facilities at Oak Ridge.[58]

Here, we shall examine the subsequent course followed by the Institute of Radiation Biology and Physicochemical Biology. In August of the following year, another large-scale international scientific conference titled The Fifth International Congress of Biochemistry was conducted in Moscow. Approximately 5,500 scientists and researchers came to this congress from 58 countries, including about 2,000 people from the Soviet Union. Totally 1,990 people read their papers, and 1,596 people expressed their questions, opinions, and comments in the discussions.[59] The papers and discussions presented at this giant congress were published in nine volumes of the proceedings.[60] The eminent physiologists, biochemists or physical chemists such as Severo Ochoa (1905–1993. A Nobel Prize laureate in Physiology or Medicine in 1959), Fritz Lipmann (1899–1986. A Nobel Prize laureate in Physiology or Medicine in 1953), Edward Tatum

(1909–1975. A Nobel Prize laureate in Physiology or Medicine in 1958), John Desmond Bernal (1901–1971), and others took part in this congress.[61] For this congress, Monod, Kalker, and others visited Moscow again.[62] The Western scientific achievements surpassed those of the Soviet science also at this congress. The report on scientific research activities of the Institute of Radiation Biology and Physicochemical Biology says,

> For 1960–61, in this area, a number of large, sometimes outstanding successes have been achieved abroad. Many of the latest research results were the subjects of presentations at the 5th International Biochemical Congress of Biochemistry held in Moscow this summer, where one of the symposia was entirely devoted to the problems of molecular biology. . . . However, even some individual successes of the Soviet science in the field of studies on nucleic acids and protein cannot compensate and hide the obvious lag in all the front of research from foreign science and the level dictated by the needs of the country.[63]

What was more serious with this institute was the difficulty in preparing the research equipment and the conditions. The Institute's report on scientific research activities for 1962 complains,

> Despite all the efforts of the Institute's leadership, it is not possible to resolve the problem of building a block for various radiation sources, a specialized building . . . The absence of such radiation sources, including sources of neutrons, α-rays, heavy nuclei, hampers the development of radiobiological studies at the Institute. In the reporting year, a small group of X-ray sources was created, but by its nature it will not be able to meet the complex requirements imposed by modern radiobiology for radiation sources.[64]

Under such difficult conditions, the basic research subjects of the institute had to be shifted to fundamental research subjects of molecular biology. The Institute's report on scientific research activities for 1963 mentions such a situation,

> The efforts of the Institute's scientists are concentrated on the solution of the following main subjects: 1. Study on the mechanism of reduplication of nucleoprotein, structure, physicochemical properties and cytochemistry of DNA and DNA-proteins. 2. Study on the structure, physicochemical properties and cytochemistry of RNA and RNA-proteins in connection with their roles in protein biosynthesis. 3. Study on the structure and mechanism of nucleic acid – protein metabolism. . . . In resolving these problems, special attention is paid to the study of the effects of various types of ionizing radiation.[65]

Here, the research on radiation effects to living body was relocated from one of the major independent objectives of the Institute's research activities to the "emphases" attached to the other main research targets. Later, in 1965, the institute was renamed the Institute of Molecular Biology. It is supposed that such renaming was needed not only because the term "molecular biology" was widespread but also because the word "radiation" was short of expectations.

4 Toward the normalization of biological sciences

4.1 Reevaluation of Aleksei Severtsov, the Selected Works of Nikolai Vavilov, *and "the Letter of Three Hundreds" as preliminary skirmishes*

Even after the establishment of the predominance of Lysenkoist doctrines in biological sciences, normal genetic research was maintained and conducted in several research facilities even with only scanty research resources.[66] Gregorii Khrushchov (1897–1962), the former Director of the Institute of Cytology, Histology and Embryology of the USSR Academy of Sciences (formerly, the Institute of Experimental Biology), once tried to protect his own position by severely criticizing his colleague, Nikolai Dubinin, in his review at the tide of the Lysenkoites' offensive against Dubinin and other geneticists.[67] Nevertheless, he took actions for rehabilitation of the scientists negated by the Lysenkoites immediately after the death of Stalin. At a meeting of the Academy's Presidium held on March 27, 1953, he proposed, in a discreet way of speech, the reevaluation of an outstanding Russian biologist, Aleksei Severtsov (1866–1936), who was criticized by the Lysenkoists for his "idealist metaphysics" in his view some years before. He said,

> The great subject set before the Soviet science by the Nineteenth Congress of the Communist Party – to take the first place in world science – requires a careful assessment of the path traversed by the Soviet science and a clear definition of the urgent and prospective subjects in each field of knowledge . . . At present, every important and every little field of biology faces the urgent subject of revising theoretical attitudes in the light of the Soviet creative Darwinism. The direction developed in the works of Academician A. N. Severtsov, that is called 'the evolutionary morphology of A.N. Severtsov,' precisely belongs to the systems of views and theoretical generalizations in the field of animal morphology. Nowadays, the views of A.N. Severtsov have been widespread and widely promoted. The interest to his work is significantly animated again. All his theories are formalistic, because there is a form in his theories and it changes by itself, independently of its content.[68]

Severtsov, a biologist who played an active role in Imperial Russia, was regarded as one of the introducers of Darwin's theory of evolution in Russia and one of the creators of a new branch of science, i.e., evolutionary morphology. In the Soviet era, he was considered as one of the outstanding scientists who "developed and protected Darwinism from the reactionary attacks from a number of West European biologists-idealists."[69] However, as the Lysenkoists began to criticize Ivan Ivan Shmal'gauzen (1884–1961), who was generally admitted to be a successor to

Severtsov and had been in the post of the Director of the Institute of Evolutionary Morphology named after A.N. Severtsov until 1949, they also criticized Severtsov on his "idealistically metaphysical" view so as to suit their own convenience for the anti-Shmal'gauzen campaign. In the meeting of the Presidium on March 27, in opposition to Khrushchov's proposal, V.P Gagarin, a Lysenkoite, asserted that the evaluation of Severtsov was confirmed by a large-scale special scientific conference called by the Minister of Higher Education and held in a lecture hall of Moscow State University in a gathering of 15 biologists, 15 philosophers, and approximately 500 audience members.[70] Therefore, a new conference for the reevaluation of Severtsov, titled "Conference on Critical Evaluation on A.N. Severtsov's *Evolyutsionnaya Morfologiya* (*Evolutionary Morphology*) and the Subjects of Animal Morphology," had to be organized by the Presidium of the Academy of Sciences. Khrushchov took the platform in the meeting of March 27 in his capacity as a deputy chair of the Organizing Committee for that conference.[71]

Eventually, this conference was held on April 24 and 25. The conclusions of this conference were confirmed by the Academy's Presidium in its meeting on May 22. Its resolution restored Severtsov's honor, blaming only Abram Deborin (1881–1963), an old philosopher, saying, "The first philosophical evaluation on the views of A.N. Severtsov, which initiated the understanding of Severtsov as a spontaneous dialectic materialist, was proposed by A.M. Deborin not from the standpoint of Marxism but from the erroneous position of Menshevik idealism."[72] This incident was the first counterattack of the scientists enrolled in the Academy of Sciences against the Lysenkoites.

Later, in a Presidium meeting on January 13, 1955, Grigorii Khrushchov proposed the Presidium to confirm the conclusion of the Division of Biological Sciences to edit and publish the *Selected Works of Nikolai Vavilov* (1887–1943).[73] Even Before the restoration of a prominent geneticist who was arrested and shot to death, Nikolai Vavilov's honor was sentenced by the Military Board of the Supreme Court of the Soviet Union later in August 20 of that year, an expectation for his rehabilitation already grew among the Academy's scientists. What is interesting here is that there was no objection to the proposal of Khrushchov in that Presidium meeting. This proposal was eventually adopted in a meeting of January 13, 1956, about one year after the proposal.[74]

A total of 94 scientists, including 70 biologists such as Dubinin, Iosif Rapoport (1912–1990), and Anton Zhebrak (1901–1965), and 24 scientists in other fields like Tamm, Kapitsa, Artsimovich, Lev Landau (1908–1968. A Nobel laureate in Physics in 1962), Andrei Sakharov (1921–1989. A Nobel Peace Prize laureate in 1975), Yakov Zel'dovich (1914–1987), Vitalii Ginzburg (1916–2009. A Nobel laureate in Physics in 2003), and others, sent a letter dated October 11, 1955, to appeal for the normalization of biological sciences in their joint signature to the Party's Central Committee. In February of the following year, the 203 additional signatures of scientists including some influential mathematicians such as Mikhail Lavrent'ev (1900–1980), Mstislav Keldysh (1911–1978), Sergei Sobolev (1908–1989), Aleksei Lyapunov (1911–1973), and others (some of them were the leaders of "Cybernetics" movement) were sent to the Party's Central Committee.

This letter came to be called, "the Letter of Three Hundreds (Pis'mo Tryokh Sot)."[75] The scientists who were apprehensive of the Lysenkoite predominance and were eager for the normalization of biological sciences succeeded in obtaining the supports from many scientists of various fields beyond the differences of their specialties, including those influential physicists and mathematicians who remarkably raised their prestige in the Soviet society by their contributions to the nuclear weapon developments and other Cold War scientific projects achieved in a very short period. At this point, it is said that this incident, the so-called "the Letter of Three Hundreds," revealed a new shape of social movements by scientists. Nevertheless, this letter did not bear any fruit. Far from it, this letter remained unknown to the public until it was reported in the Party's organ, *Pravda*, in 1987.[76]

4.2 "Nikolai Dubinin's Institute"

Besides the Institute of Radiation Biology and Physicochemical Biology, another large scientific research center was to be established in the Siberian Branch of the Soviet Academy of Sciences. Its foundation had been already decided at the Presidium meeting on June 21, 1957. At the same time, Nikolai Dubinin was appointed as its director-organizer.[77] He was given a "carte blanche" in selection of staff and choice of research subjects for the new institute. Dubinin recognized well that the new institute with himself as Director was expected to contribute to a rapid development of radiation genetics, saying,[78]

> Now as the important subjects are raised in connection with the organization of the Institute of Radiation Biology [correctly – the Institute for Radiation Biology and Physicochemical Biology. . . *Ichikawa*], and so the work on radiation genetics begins to be expanded, which is of great importance for the future of the institute.

According to an address in July 1957 by the main organizer of the Siberian Branch Mikhail Lavrent'ev at a general assembly of the Academy of Sciences on July 2–6, 1957, convened to confirm the above-mentioned decision of the Presidium on June 21, the Institute of Cytology and Genetics of the Siberian Branch was to hire more than 400 persons in the vast space of 5,000 m^2.[79]

Apparently Dubinin and other anti-Lysenko biologists, however, aimed at the rapid development of wide range of new trends in biology. In the Western countries molecular biological study rapidly developed especially since the publication of a paper by James D. Watson (1928–. A Nobel laureate in Physiology or Medicine, 1962) and Francis H.C. Crick (1916–2004. A Nobel laureate with J. Watson) in 1953, which identified the double helical structure of Deoxyribonucleic acid. Soviet biologists could not just look on that boom with their arms folded. Molecular biology making full use of physical and chemical methods is useful not only for radiation research but also the radical change of the situation in the Soviet biology dominated by "the monopoly of the truth" by the Lysenkoite's doctrine rejecting "material bases of heredity." In order to catch up soon with the level of molecular biology in the West, a rapid development of a wide range of fundamental research subjects was needed. The scale of his institute with a staff of almost a hundred persons even in its very early days could not satisfy Dubinin.[80] He complained at the first general assembly of the Siberian Branch held on May 15–19, 1958; "It seems to me that we cannot be satisfied with the existing organization of works on biology." He continues,[81]

> The Presidium of the Siberian Branch, making a progress in the task of creating the profiled biological institutions, exchanged the opinions with the Department for Biological Sciences of the Academy of Sciences of the USSR. Eventually the Department issued the proclamations to build ten new biological institutes in Siberia. . . . Todays, the problem of heredity, and then, biology as a whole are going through a remarkable period. Molecular bases of heredity were revealed as nucleic acid and there is every reason to believe that the nucleic acid is the compound related primarily with the phenomenon of heredity.

The structure of his institute reflected such an overall interest covering every field of biology. According to the report[82] dated December 12, 1958, the institute consisted of totally six divisions and 23 laboratories: (1) Within the Division of Physical, Chemical and Cytological Basics of Heredity – totally five laboratories – 1, Laboratory of Cytological Basics of Heredity; 2, Laboratory of Nucleic Acid and Nucleoprotein; 3, Laboratory of Micro-Spectrophotometry and Electronic Microscopy; 4, Laboratory of Cell Embryology of Plants; 5, Laboratory of Scientific Micro-Cinematography. (2) Within the Division of General and Radiation Genetics – totally six laboratories – 1, Laboratory of General Genetics; 2, Laboratory of Population Genetics; 3, Laboratory of Radiation Genetics; 4, Laboratory of Cytogenetics; 5, Laboratory of Polyploidy of Plants and Animals; 6, Laboratory of General Genetic Methods for Selective Breeding. (3) Within the Division of Genetics and Cytology of Plants – totally three laboratories – 1, Laboratory of Radiation Breeding and Experiential Mutations; 2, Laboratory of Heterosis and Hybridization; 3, Laboratory of Cytology and Apomixes. (4) Within the Division of Animal Genetics – totally four laboratories – 1, Laboratory of Individual Genetics and Cytology of the Animals; 2, Laboratory of Hybridization and Heterosis in Animals; 3, Laboratory genetic foundations of breeding animals; 4, Laboratory of Ecological Genetics of Animals. (5) Within the Division of Genetics and Cytology of Cancer – totally two laboratories – 1, Laboratory of Genetics of Cancer; 2, Laboratory of Cytology of Cancer Cells. (6) Within the Division of Genetics and Cytology of Microorganisms and Viruses – totally three laboratories – 1, Laboratory of Genetics and Cytology, Electronic Viruses and Phages; 2, Laboratory of Cytogenetics Fungus; 3, Laboratory of Genetics and Cytology of Bacteria.

4.3 Harassments of the Lysenkoites

The persistent harassment and interferences by the Lysenkoites against the anti-Lysenko biologists are already well known by the witnesses of the involved persons.[83] Even after his dismissal from the post of President of All-Union Academy of Agricultural Sciences named after V.I. Lenin (Hereafter, VASKhNIL), Lysenko did not stop urging political authorities to restrain their opponents from gaining power in biological sciences.[84]

Immediately after his group restored its relationship with high political authorities in the spring in 1957, Trofim Lysenko (1898–1976) sent a letter dated April 19 to Aleksandr Nesmeyanov, President of the Academy. In it, he condemned "the situation in which scientific research activities that are carried out or are to be carried out with the application of isotopes and nuclear radiation, I don't know why, are now conducted by outsiders of the Institute of Genetics." He demanded that Nesmeyanov should "remove the fictitious leaders away."[85] He also said: "In the Institute of Genetics, genetic studies on radiation effects are well organized. They are being carried out on the basis of Michurinist biology."[86] On October 8–14, his institute held a large-scale scientific conference, gathering a total of 375 participants from 250 scientific organizations and higher education facilities. In it, Nikolai Nuzhdin (1904–1972), Lysenko's right-hand man, delivered a speech, proclaiming, "The chromosome theory and the gene hypothesis turned out to be bitterly contrary to the materialist view."[87] As a maneuver, or perhaps in triumph, the conference organizers invited a number of anti-Lysenko biologists such as Dubinin and Shmal'gauzen, their sympathizer Kapitsa, some philosophers including Mark Mitin (1901–1987), Aleksandr Maksimov (1890–1976), and Abram Deborin among others.[88]

A corresponding member of VASKhNIL, Professor I.S. Sizov from All-Union Institute of Plant Breeding, sent his letter dated August 7, 1959, to the editorial board of the Party's organ, *Pravda*. In this letter Sizov expressed his doubts about the reason for establishing the Institute of Cytology and Genetics of the Siberian Branch of the Academy of Sciences, saying,[89]

> Is it appropriate to disperse in such a way the efforts on genetic and cytogenetic studies. Would it be better to concentrate all the work in the Institute of Genetics of the USSR Academy of Sciences? There arose an impression that some individual workers of the Biological Department of the USSR Academy of Sciences apparently didn't like the Michurinist direction in genetics conducted by T.D. Lysenko and then the second Institute was established in the opposite direction in science in Novosibilsk.

Having received his letter, the editorial board of *Pravda* referred to Lavrent'ev, Director of the Siberian Branch, for the answer to Sizov. Lavrent'ev sent his reply dated September 25 directly to Sizov. In his reply in the first place Lavrent'ev called Sizov's attention to the particular position of the Siberian Branch that unlike other local organization (filial) in the Siberian Branch any research subject is not restricted to the local issues. And then, he evaded Sizov's question with the help of formal logic on the procedure, saying,[90]

> The final decisions on the structures and the directors of the institutes are adopted on the recommendation of the related department and the general assemblies of the Academy of Sciences of the USSR and approved by the Government. The biologists and the experts in the related fields of science believed the establishment of the Institute of Cytology and Genetics appropriate.

4.4 *"Commission on Biology" of Central Committee of the Party: pluralism in scientific research*

Lysenko could not look on unconcernedly at the rapid expansion of "Dubinin's Institute" in Siberia. He wrote to Nikita Khrushchev and demanded him to interfere in biological sciences. Eventually a commission composed of some leading Lysenkoites like A.G. Utekhin (then, Head of the Scientific Section of the Department of Agriculture, the Party's Central Committee), Mikhail Ol'shanskii (1908–1988, later, the USSR Minister of Agriculture in 1960–1962), Nikolai Nuzhdin, and others was set up under the name of "Commission on Biology" by the Party's Central Committee in January 1959. This commission was dispatched to the academic city of Akademgorodok near Novosibirsk in order to interrogate the leadership of the Siberian Branch on the matter related with biological sciences. That mission visited some Bureau members of the Presidium of the Siberian Branch like Lavrent'ev, Timofei Gorbachyov (1900–1973, a metallurgist. Then, Vice Director of the Siberin Branch in 1957–1972), Andrei Trofimuk (1911–1999, a geologist), Dubinin himself, and others. Here, let's take a look at the stenographic note of their meeting, which was kept in the Scientific Archive of the Siberian Branch.[91]

As soon as the meeting discussed the main subject, Ol'shanskii said,

> The Commission considers the genetic direction not much successful and it is not fruitful to teach young people in detail that there are two approaches [Lysenkoites' "Michurinisit agronomy" and the molecular biological methods of genetics approved by Dubinin and others – *Ichikawa*] to the solution of biological issues, as confirmed officially. And the Commission does not share the approach seen in and around this Institute as a principle. . . . It is not proper to entrust biology exclusively to only one possible direction [the molecular biological methods of genetics – *Ichikawa*]. Such a direction cannot be considered sufficiently promising to further progress of science. [Sheet No. 3]

The oddest thing here is that the Lysenkoites, very often considered to be seekers of "monopoly of scientific truth," demanded breakdown of monopoly in scientific research, shutting their own eyes to their own behaviors. Lavrent'ev quickly asked back, "Does that mean – strengthening the other approach of science ['Michurinist biology' – *Ichikawa*]?" [Sheet No. 3] Ol'shanskii soon replied, "Yes." [Sheet No. 3] Lavrent'ev's following remark was a sort of maneuver: "Here, the point is that here is very small and we do not have yet the necessary space. . . . It is needed to help us so as that the persons having another direction can smoothly settle in their works after coming here." [Sheet No. 4] Gorbachyov covered Lavrent'ev: "All the doctors you recommended, we adopted." [Sheet No. 4]

No sooner than Nuzhdin remarked, "There are two directions. And I have some sympathy for each of them," [Sheet No. 5] Trofimuk replied, "But each of them has different labels." [Sheet No. 5] The Siberian Branch member showed a feeling of antipathy against the idea, "pluralism of scientific research," raised by the Lysenkoites. Lavrent'ev shows a sentiment, remembering the formidable days at the beginning of the Siberian Branch of the Academy of Sciences:

> The establishment of a scientific center in Novosibirsk had been considered as the most harmful attempt in many ways. The young persons were persuaded not to come here. We had great difficulty in securing the necessary persons. I would like to frankly tell you when A. N. Nesmeyanov told me about the conversation with Nikita Sergeivich. He said, "No one will go to you." A. V. Topchiev was also present at this conversation. This conversation worried us so much. Having taken the advantage of this conversation, you censured us, as if you fingered a wart. [Sheet No. 6]

A plant philologist, Pavel Genkel' (1903–1985), a Commission member, summarized their demands:

> No one is prepared to remove the director from his position. You have to work. Our task was not to make decisions. We had to be familiar with the directions [of the Institute of Cytology and Genetics – *Ichikawa*]. These directions go against the Michurinist directions. There are working here 112 people (some in Moscow and Khar'kov). However there is no aggressive comrades standing on Michurinist direction. . . .[92] It does not sound good, although not the worst, that Nikolai Petrovich [Dubinin – *Ichikawa*] is concentrating all his efforts to only one direction. [Sheet No. 10]

Utekhin covered Genkel': "We therefore call on the workers in the Institute of Cytology and Genetics more quickly obtain the necessary production results." [Sheet No. 10] They meant that they would like to share the new institute in the Siberian Branch by obtaining some stage for their activities there and, if the institute had allowed them to develop the methods for improvement of productivity in agriculture, based on "the Michurinist methods," they should not request the dismissal of Dubinin from the post of Director of the institute.

Ol'shanskii was, however, more radical. He demanded an overall change of direction of the institute: "I said that the direction of the Institute stands on a methodologically incorrect positions, including all the results caused by that direction." [Sheet No. 10] In opposition to this, Lavrent'ev emphasized diversity of the scientific methods; "What does it mean 'methodologically incorrect?' . . . The sugar beet – to increase the output of sugar, are they approaching correctly to this issue? Or it is possible to do it another way, isn't it?" [Sheets Nos.10,11]

Ol'shanskii replied, "We cannot say that crap what is made by idealists is all absurd. But, it is needless to say, the science based on the materialist method certainly would be more fruitful." [Sheet No. 11] After Lavrent'ev's further persuasion, Ol'shanskii at last answered in affirmation: "We don't deny none of the ways that benefits." [Sheet No. 11] And then he kept silent after this remark.

Dubinin tried to put an end to this theatrical dialogue: "I am pleased that Comrade Utekhin does not support the view point of Academician Ol'shanskii, i.e., the opinion that the Institute should be closed. It means that our work must continue. And all your comments we will consider." [Sheet No. 12] Lavrent'ev covered him: "After the 21st [Party's – *Ichikawa*] Convention we will reconsider all thematic plans." Utekhin made a concluding remark, again emphasizing "pluralism in scientific research," saying, "The Institute needs to take four or five major subjects: one for Director, one for his Deputy and the others are to be shared among the Departments heads and has to make the results of the works observed and put these tasks under daily control." [Sheet No. 12]

4.5 Toward the downfall of the Lysenkoism

If we see the downfall of the Lysenkoites' hegemony as a necessary result of "de-Stalinization of the Soviet science" in general, this seems somewhat simplistic. We must take into consideration the context in which the research on "radiation effect on living bodies" came to obtain major significance in the Soviet Union in those days. As the nuclear weapons race between the Soviet Union and the United States intensified from the mid-1950s, that research had to become the main research target for both "Super Powers." This relationship between science and the Cold War was a self-evident truth for the physicists involved in any nuclear research project, whether it may be of military purposes or peaceful purposes.

The minutes and the stenographical records of the Presidium meetings tell us how serious and important leading scientists regarded the radiation study. They expected anti-Lysenkoite geneticists and biologists to make rapid progress in this field.[93]

At the same time, it must be taken into consideration that almost simultaneously, molecular biology developed rapidly across the world. Dubinin and other biologists were pressured to make a breakthrough for catching up with the rest of the world in a wide range of basic research in this new field of science. "The Dubinin's Institute" thus became a huge-scale scientific organization treating a big variety of research subjects related with molecular biology, not limited to the radiation effect study. Such an expansion of research subjects makes the original purpose of the establishment of the Institute (rapid development of radiation effect study) somewhat hard to see. Eventually molecular biology was placed in the Party's new Program adopted at the 22nd Party's Convention. Subsequently, on May 11, 1962, the Academy's Presidium set up a large-scale scientific board consisting of 68 members with Engel'gardt as Chairman for the policy-making for the further development of molecular biology.[94]

Here let us see how the matter developed. On January 9, 1963, a joint meeting of the Central Committee of the Communist Party and the Council of Ministers of the USSR issued a proclamation titled "On the measures for the further development of biology and the reinforcement of its linkage with the practice."[95] Subsequently on February 1, a joint conference of the Presidium and the Department of Biological Sciences of the Academy of Sciences was held. In its resolution the joint conference evaluated highly "the Michurinist direction" in biology and adopted totally 135 research subjects most of which were based on the Michurinist agronomy. Six of them were to be carried out by Lysenko himself and his close comrades.[96] The representatives from various research institutions reported about the positive results by application of "the Michurinisit directions" on the platform one after the other for four hours during the conference.[97] In his concluding remark, President of the Academy Mstislav Keldysh (President in 1961–1975) emphasized the importance of "pluralism in scientific study," saying, "It seems to me that the discussion here vividly demonstrated the more diversified are the objects in the living world which we make use of, the more diversified are the methods utilizing them and the approaches to their research."[98] That remark was all he could make to show his resistance against the "monopoly of scientific truth" by Lysenkoites. Even talkative Kapitsa kept his silence for four hours. Dubinin also did so. Another person who kept silent was Lysenko himself.[99] The downfall of Lysenkoites' hegemony was, then, not so straightforward.

Notes

1 Arkhiv Rossiiskoi Akademii nauk (ARAN), Fond (F.) 694, Opis' (Op.) 1, Delo (D.) 101, 15.
2 V.N. Mikhailov (Ed.), *Sozdanie Pervoi Sovietskoi Yadernoi Bomby* (*Making of the First Soviet Nuclear Bomb*). Moscow, Energoatomizdat, 1995g, 154.
3 A.M. Petros'yants et al. (Ed.), *Yadernaya Industriya Rossii* (*Russian Nuclear Industry*). Moscow, Energoatomizdat, 2000g, 321, 322.
4 A.K. Kruglov, *Kak Sozdavalas' Atomnaya Promyshlennost' v SSSR* (*How Was the Atomic Industry Created in the USSR?*). Moscow, TsNII ATOMINFORM, 1995g, 72–80; From 1948 to 1952, in the Soviet Union, radiation exposure below 30 rem was considered permissive (see, *Sozdanie Pervoi Sovietskoi Yadernoi Bomby* . . . . in note 2. 116.)
5 *Sozdanie Pervoi Sovietskoi Yadernoi Bomby* . . . in note 2. 116; Pod red. Petros'yants et al. . . . in note 3. 332; It was also reported that 36–43% of all the workers were exposed to radiation equivalent of 100–400 rem (*Sozdanie Pervoi Sovietskoi Yadernoi Bomby* . . . in note 2. 116).
6 Ibid., 154.
7 Kruglov, *Kak Sozdavalos'* . . . in note 4. 101.
8 *Sozdanie Pervoi Sovietskoi Yadernoi Bomby* . . . in note 2. 161, 162.
9 Ibid., 147.
10 Ibid., 159.
11 Institut Promyshlennoi Ekologii Ural'skogo Otdeleniya Rossiiskoi Akademii Nauk [pod red. V.N. Chukanova], *Vostochno-Ural'skii Radioaktivnyi Sred: Sverdrovskaya Oblast'* [Institute of Industrial Ecology, the Uralian Branch of Russian

Academy of Sciences (Ed. by V.N. Chukanov), *The East Uralian Radioactive Trace: Sverdrov Oblast*]. Ekaterinburg, 1996.

12 Institut Promyshlennoi Ekologii Ural'skogo Otdeleniya Rossiiskoi Akademii Nauk [pod red. V.N. Chukanova], *Vostochno-Ural'skii Radioaktivnyi Sred: Problem Reabilitatsii Naseleniya v Territorii Sverdrovskoi Oblasti* [Institute of Industrial Ecology, the Uralian Branch of Russian Academy of Sciences (Ed. by V.N. Chukanov), *The East Uralian Radioactive Trace: The Problems of Rehabilitation in the Territories of Sverdrov Oblast*]. Ekaterinburg, 2000: Zhores Medvedev once said, "Despite the tragic nature of the disaster, the existence of such an extensive zone of contamination containing radioactive materials at different levels of concentration offered a unique opportunity for scientific research in such areas as radioecology, radiogenetics, radiobiology, radiotoxicology." (Zhores Medvedev, trans. by George Saunders, *Nuclear Disaster in the Urals*. New York, W.W. Norton & Company, 1979, 22). This remark by Medvedev, however, has been not supported by any official document of the Soviet and Russian authorities.

13 "Valentin Fedorovich Khokhryakov: K 70-letiyu so dnya rozhdeniya." *Voprosy Radiatsionnoi Bezopasnosti*, No. 2, 1998, 80.

14 V.F. Khokhryakov i dr., "Vnutrennee obuluchenie personara PO 'Mayak'." *Voprosy Radiatsionnoi Bezopasnosti*, No. 3, 2000, 51–53.

15 V.V. Khokhryakov, V.F. Khokhryakov i dr., "Anariz dinamiki izotopnogo sostova aktinidov, postupavshikh v organizm rabotnikov PO 'Mayak'." *Voprosy Radiatsionnoi Bezopasnosti*, No. 3, 2007, 58, 59, 64, 65; After the collapse of the Soviet Union, they came to engage themselves in the large-scale and serial investigations to clarify the dynamism of the effects of the actinide isotopes and the radioactive nuclides in the human organs and tissues on the basis of the measurement of α-ray from the bone tissues gained at the postmortem examinations of the former workers who had died from 1973 to 1993.

16 L.A. Orbeli, "Deistvie radiatsii na organism zhivotnykh." in Akademiya Nauk SSSR (Ed.), *Sessiya po Mirnomu Ispol'zovaniyu Atomnoi Energii* [1–5 Iyulya 1955g], Vol. 2. Moskva, Izd-vo AN SSSR, 1955g, 15.

17 P.E. Minaev, "O lokal'nom deistvii rentgenovskikh luchei na razlichnye otdely tsentral'noi nervnoi sistemy zhivotnykh." ibid., 51–67.

18 Orbeli, "Deistvie radiatsii . . ." in note 16. 10–11.

19 A.G. Pasynskii, "Deistvie ioniziruyushchei radiatsii na rastvory belkov i belkovykh kompleksov." ibid., 85–105; This kind of ionization and oxidization had been already reported by Nikolai Timofeev-Ressovsky (1900–1981) and others as early as mid-1930's.

20 Orbeli, "Deistvie Radiatsii . . ." in note 16. 6; A.M. Kuzhin, "Biokhimicheskie Osnovy Biologicheskogo Deistviya Ioniziruyushchei Radiatsii." ibid., 69–82.

21 "Pervaya Vsesoyuznaya Konferentsiya po Meditsinskoi Radiologii." *Atomnaya Energiya*, Vol. 1, No. 2, 1956, 99.

22 M.N. Livanov, "Izmeneniya, nastupayushchie v razlichnykh otdelakh tsentral'noi nervnoi sistemy u zhivotnykh posle vozdeistviya luchei rentgena." in Ministerstvo zdravookhraneniya SSSR [A.V. Lebedinskii – otvetstvennyi redaktor] (Ed.), *Tezisy Sektsionnykh Dokladov Vsesoyuznoi Konferentsii po Meditsinskoi Radiologii: Sektsiya Eksperimental'noi Radiologii*. Moskva, Medgiz, 1956, 3–6.

23 A.A. Bagdasarov and L.S. Rogachev, "Lechenie ostroi i khronicheskoi formy luchevoi bolezni." in Ministerstvo zdravookhraneniya SSSR [A.V. Lebedinskii – otvetstvennyi redactor] (Ed.), *Tezisy Sektsionnykh Dokladov Vsesoyuznoi Konferentsii po Meditsinskoi Radiologii: Klinicheskaya Sektsiya*. Moskva, Medgiz, 1956, 17, 18.

24 A.A. Letaved, "Gigienicheskie problemy v radiologii." in Ministerstvo zdravookhraneniya SSSR [A.V. Lebedinskii – otvetstvennyi redactor] (Ed.), *Tezisy Plenarnykh Dokladov Vsesoyuznoi Konferentsii po Meditsinskoi Rradiologii*. Moskva, Medgiz, 1956, 27, 28.

25 ARAN, F. 2, Op. 6, D. 244, 179, 193.
26 V.P. Shvedov and L.I. Gedeonov, "Zagryaznenie biosfery v okrestnostyakh Leningrada produktami yadernykh vzryvov." Pod otvet. red., A.V. Kozlovy, *Meditsinskaya Radiologiya: Trudy Vsesoyuznoi Nauchno-Tekhnicheskoi Konferentsii po Primeneniyu Radioaktivnykh i Stavil'nykh Izotopov i Izluchenii v Narodnom Khozyaistve i Nauke* [4–12 Aprelya 1957g]. Moskva, Izd-vo AN SSSR, 1960, 45–57.
27 V.M. Klechkovskii and I.V. Guryakin, "Postuplenie v rasteniya radioaktivnogo strontsiya i nakoplenie ego v urozhae razlichnykh sel'skokhozyaistvennykh kul'tur." ibid., 58–65.
28 G.G. Tinyakov and M.A. Arsen'eva, "Tsitogeneticheskie posledstviya vozdeistviya radiatsii na spermatogenez obez'yan." ibid., 100–117.
29 M.N. Pobedinskii, *Luchevaya Bolezn'*. Moskva, Medgiz, 1957.
30 As early as November 10, 1949, the USSR representative to the United Nations, Andrei Vyshinskii (1883–1954), referred to the non-military use of atomic energy for the first time in the international arena of the Fourth General Assembly of the United Nations (Cited in V. Vovulenko, "Vstpitel'naya stat'ya (Introductory Article)." in Dzh. Allen, *Atomnaya energiya i obshchestvo* [A Russian version of James Allen, *Atomic Energy and Society*. New York, International Publishers, 1949]. Moskva, Izdatel'stvo innostrannoi literatury (Foreign Literature Publisher), 1950, 5–19.). Furthermore, on October 5, 1952, a year and two months earlier than the "Atoms for Peace" address made by US President Dwight Eisenhower (1890–1969), a Politburo member, Georgii Malen'kov (1902–1988), spoke highly of the peaceful use of atomic energy at the first day of the meeting of the Nineteenth Convention of the All-Union Communist Party (Bol'sheviki) (G.M. Malen'kov, "Ochyotnyi doklad Tsentral'nogo komiteta VKP(b) XIX s"ezdu partii." http://stalinism.ru/dokumentyi/materialy-xix-s-ezda-vkp-b-kpss.html?tmpl=component&print =1&page=). After this Convention, a brilliant future of atomic energy became one of the main targets of propaganda in the Soviet press; an article titled "Atomic Energy for Peaceful Purposes" appeared in the most influential scientific enlightening journals, *Znanie – Sila* (*Knowledge Is Power*) by a chemist, A. Seregin (A. Seregin, "Atomnaya energiya dlya mirnykh tselei." *Znanie – sila*, No. 3, 1953g, 27, 28). On May of the following year, a journal of Komsomol (Communist Youth League), *Tekhnika Molodezhi* (*Technology for Young Persons*), carried an article, "Nuclear Reactors" by a famous scientific writer-popularizer, K. Gladkov (K. Gladkov, "Yadernye reaktory." *Tekhnika Molodezhi*, No. 5, 1954, 23–29.) As for the reality of the initial Soviet nuclear development for the peaceful purpose, see Chapter 6 of this book.
31 Akademiya Nauk SSSR, *Sessiya po Mirnomu Ispol'zovaniyu Atomnoi Energii* (*The Session on the Peaceful Use of Atomic Energy*), 1–5 Iyulya 1955g, Vol. 1–5. Moskva, Izd-vo AN SSSR, 1955.
32 Vol. 2 of the five-volume proceedings of this Session is dedicated to the biological problems of radiation.
33 ARAN, F. 2, Op. 6, No. 201, 138–140. The remark by Samarin can be seen in Sheet No. 140.
34 ARAN, F. 2, Op. 6, No. 221, 30–96: Nikolai Dubinin was appointed as Chief of this laboratory. In 1957, in the Institute of Chemical Physics of which the Director was Nikolai Semyonov (1896–1986. A Nobel laureate in Chemistry in 1956), a new department, Department of Biochemistry, was established, and Iosif Rapoport (1912–1990), who had been long purged by Lysenkoites, was invited as Department Head (Varelii Soifer, *Vlast' i Nauka: Razgrom Kommunistami Genetiki v SSSR*. Moskva, CheRo, 2002, 850). Concerning Dubinin's life and activities, see, A.A. Zhuchenko and L.K. Ernsta (Ed.), *Nikolai Petrovich Dubinin i XX vek: Sovremenniki o Zhizni i Deyztel'nosti: Pis'ma, Materialy, Vospominaniya* (sost. – L.G. Dubinina and I.N. Ovchinnikova). Moscow,

Nauka, 2006g; *K Stoletiyu Akademika N.P. Dubinina: Po Materialam Nauchnoi Konferentsii "Genetika na Rubezhe Vekov"*, Moskva, 22–23 marta 2007 goda. Moskva, Nauka, 2009. and others.

35 See, Nakagawa Yasuo, *Hôshasenn Hibaku no Rekishi: Amerika Genbaku Kaihatsu kara Fukusima Jiko made* (*A History of Radiation Exposure: from American A-Bomb Development to Fukushima Accident*). Tokyo, Akashi Shoten, 2011, 70–90.

36 ARAN, F. 2, Op. 6, No. 263, 32: This conference in Paris had totally 1,171 participants. They heard 206 presentations, of which 49 were presented by the Soviet, Ukrainian, and Belorussian participants. That number exceeded those from France (33), the United States (31), and the United Kingdom (29). The proceedings of this conference were published as, R.C. Extermann (Ed.), "Radioisotopes in Scientific Research." *Proceedings of the International Conference held in Paris in September 1957*, Vol. 4. London, Pergamon Press, 1958. Preceding this, a large-scale conference, "All-Union Scientific and Technological Conference on the Application of Radioactive Isotopes and Radiation" was held jointly by the Academy of Sciences and "Main Directorate on Utilization of Atomic Energy [correctness of that name is somewhat doubtful – *Ichikawa*]." Approximately 3,000 participants gathered from totally 1,016 organizations and heard 444 presentations (ARAN, F. 2, Op. 6, No. 244, 179, 193). However, until the Paris Conference only one experimental work in laboratory scale for the industrial application of radiation for the sake of polymerization and producing graphite polymer had been put into practice (ARAN, F. 2, Op. 6, No. 263, 32).

37 ARAN, F. 2, Op. 6, No. 264, 115.

38 ARAN, F. 2, Op. 6, No. 231, 160–178.

39 ARAN, F. 2, Op. 6, No. 240, 8.

40 Ibid., 79.

41 ARAN, F. 2, Op. 6, No. 243, 58–60, 223.

42 ARAN, F. 2, Op. 6, No. 277, 10–12.

43 Ibid., 54.

44 Ibid., 94.

45 ARAN, F. 2, Op. 6, No. 268, 124, 125.

46 ARAN, F. 1979, Op. 1, D. 1, 2.

47 ARAN, F. 1979, Op. 1, D. 3, 119–121.

48 R.J.C. Harris (Ed.), *The Initial Effects of Ionizing Radiations on Cells: A Symposium Held in Moscow, October, 1960, supported by UNESCO and the IAEA and sponsored by the Academy of Sciences of the U.S.S.R.* London and New York, Academic Press, 1961, v–xiii.

49 Ibid., 329.

50 Ibid., 331.

51 Ibid., 337.

52 Ibid., 337.

53 For example, Tumerman questioned against the Western scientists, saying, "I cannot understand why the attention of radiobiologists in particular, and at this symposium, is to such a great extent concentrated on injuries of the first type (which are restricted to the cell which has received the radiation – *Ichikawa*), and why so little attention is devoted to injuries of the second type (which are reduplicated through many cellular generations – *Ichikawa*) (ibid., 353)." Furthermore, Manoilov (his given name and patronymic – unknown) pointed out, "In Alexander's interesting paper, and also in many other communications at this symposium, it has been shown that changes in the nucleoproteins cannot be responsible for the development of radiation injury in the cells. . . . In his paper Alexander put forward the supposition of the special sensitivity of the phospholipids which enter into the composition of cellular membranes. Unfortunately

the experimental proof on behalf of this supposition were not given in the paper (ibid., 355)."

54 ARAN, F. 1979, Op. 1, D. 9, 34: This document notes that Auerbach was a British man, although he was an American.
55 Ibid., 35.
56 Ibid.
57 Richard B. Setlow, *Alexander Hollander 1898–1986: A Biographical Memoir. National Academy of Sciences*. Washington, DC, Copyright 2011, 1–9. www.nasonline.org/publications/biographical-memoirs/memoir-pdfs/hollaender-alexander.pdf.
58 W.L. Russell, L.B. Russell and E.M. Kelly, "Radiation Dose Rate and Mutation Frequency." *Nature*, Vol. 129, 1958. 1546–1550.
59 ARAN, F. 388, Op. 1, D. 520, 26.
60 N.M. Sisakian (general editor), *Proceedings of the Fifth International Congress of Biochemistry*. Oxford, London, New York, Paris, Pergamon Press, 1963.
61 Ibid., Vol. I, xi, xii; ibid., Vol. III, ix, xi.
62 Ibid., Vol. I, xi; ibid., Vol. III, x.
63 ARAN, F. 1979, Op. 1, D. 22, 2, 4.
64 ARAN, F. 1979, Op. 1, D. 35, 26.
65 ARAN, F. 1979, Op. 1, D. 50, 3.
66 Nikolai Krementsov, *Stalinist Science*. Princeton and Oxford, Princeton University Press, 1997, 105, 106.
67 Soifer, *op. cit.*, in note 34. 607.
68 ARAN, F. 2, Op. 6, D. 149, 192–194, 204.
69 ARAN, F. 2, Op. 3-a, D. 92, 156.
70 ARAN, F. 2, Op. 6, D. 149, 311.
71 Ibid., 189.
72 ARAN, F. 2, Op. 6, D. 152, 12.
73 ARAN, F. 2, Op. 6, D. 192, 254.
74 ARAN, F. 2, Op. 6, D. 210, 26.
75 "Pis'mo Tryokh Sot." in A.F. Kiselyov and E.M. Shchagin (Ed.), *Khrestomatiya po Otechestvennoi Istorii (1946–1995)*. Moscow, Izd-vo 'Vlados', 1996, 458–460.
76 *Pravda*, January 27, 1989.
77 ARAN, F. 2, Op. 6, No. 248, 9.
78 Ibid., 124.
79 Nauchnyi Arkhiv Sibirskogo Otdeleniya Rossiiskoi Akademii Nauk (The Scientific Archive of the Siberian Branch of Russian Academy of Sciences. Hereafter NA SO RAN), F. 4, Op. 1, No. 6, 27.
80 In the beginning of 1958, 101 staff members were enrolled in the institute, which was 400 m^2 in area. And the scale of the institute was planned to be doubled in the following year (NA SO RAN, F. 10, Op. 3, No. 6, 9, 10).
81 Ibid., 171, 172.
82 NA SO RAN, F. 10, Op. 3, No. 20, 9, 10.
83 See, M.A. Lavrent'ev, "Opyty zhizni: 50 let d nauke," in *Vek Lavrent'eva*. Novosibirsk, Izdatel'stvo SO RAN, 2000, 149–152; N.P. Dubinin, "Odna iz yarchaishikh stranits v moei zhizni," ibid., 228–229. Paul Josephson also refers to the difficulties in the establishment of the Institute for Cytology and Genetics in the Siberian Branch. See, Paul Josephson, *New Atlantis Revisited: Akademgorodok, The Siberian City of Science*. Princeton and Oxford, Princeton University Press, 1997, 82–119.
84 For example, Lysenko protested against the personnel change in the editorial board of the bulletin of the Academy's Department of Biological Sciences, having sent his letter dated June 14, 1956, to the Chief of the Section of Science of

Party's Central Committee, Vladimir Kirilin (1913–1999), and, subsequently, another letter dated July 5 to a Secretary of the Central Committee, Mikhail Suslov (1902–1982). And then, on November 21, the scientific council of the Academy's Institute of Genetics criticized a scientific paper by a cytologist and microbiologist, Dmitrii Petrov (1909–1987), titled "To the Problems on Material Basis of Heredity" published in the journal, *Obshchaya Biologiya* (*General Biology*), and made a decision to send the materials to the Typographical Office of the Party' Central Committee (ARAN, F. 201, Op. 1, No. 279, 38–40, 49).

85 ARAN, F. 201, Op. 1, No. 284, 8: The contents of this letter was recorded here.

86 ARAN, F. 201, Op. 1, No. 307, 16.

87 ARAN, F. 201, Op. 1, No. 305, 1–3.

88 Ibid., 38, 39.

89 NA SO RAN, F. 4 Op. 1, No. 62, 24, 25.

90 NA SO RAN, F. 4, Op. 1, No. 62, 21.

91 NA SO RAN, F. 10, Op. 3, No. 59, 1–13. Hereafter, concerning the citations from this stenographic record, only the sheet number is indicated after each remark in brackets. Concerning to this mission, also see, Lavrent'ev, *op. cit.*, in note 86. 150; Dubinin, *op. cit.*, in note 86. 228–229.

92 This remark is contradictory to the former remark by Timofei Gorbachyov, "All the doctors you recommended, we adopted."

93 In spite of such efforts, the Presidium of the Academy of Sciences had to again point out the serious delay in radiation biology (ARAN, F. 2, Op. 6a, No. 198, 8, 9).

94 ARAN, F. 2, Op. 6a, No. 197, 13, 21.

95 ARAN, F. 2, Op. 6a, No. 204, 10.

96 Ibid., 10, 27–83.

97 ARAN, F. 2, Op. 6, No. 434, 159–302.

98 ARAN, F. 2, Op. 6, No. 434, 298.

99 Ibid., 159–302.

8 Anna Vasil'evna Kozlova (1906–1980)

The fate of the data on the casualties of Hiroshima, Nagasaki, and Bikini

The eminent radiologist Anna Kozlova (1906–1980) visited Japan before the normalization of Soviet-Japanese diplomatic relations and brought back with her to the Soviet Union large amounts of data on radiation exposure from the Hiroshima, Nagasaki, and Bikini Atoll bombings. Herein, the author discusses some little-known facts about Dr. Kozlova, including her biography, her stance on the problem of radiation, and how she regarded the data that she brought back with her, with a view to elucidating how the Japanese data on radiation exposure were treated, including the data on the direct radioactive effects of nuclear bomb explosions and those of the effects of subsequent radiation exposure in Hiroshima, Nagasaki, and Bikini Atoll.

I am neither a medical doctor nor scientist but simply a Japanese woman who longs for world peace, and it is as such a woman that I take a great interest in, and pray for, the success of the international round-table on radiation. I was delighted to hear that you, only one female medical scientist, are coming from the Soviet Union to attend the conference, and I pin my high hopes on you.

(Raichô Hiratsuka, "Letter to Anna Kozlova,"
in *Complete Works*, Vol. 7, p. 300)

1 Anna Vasil'evna Kozlova

1.1 An authority on the development of radiation therapy

Anna Vasil'evna Kozlova was born on September 8, 1906, the daughter of an office clerk in St. Petersburg. After earning her medical degree from Voronezh State University in 1927, she spent some years in medical practice in various places before serving as a principal physician, and subsequently chief physician, in the Radiology (X-ray) Department of the Tashkent Cancer Research Center. It was there that she wrote her doctoral thesis, "Lip Cancer and Its Treatment." In 1945, she took up a post as a radiologist in the X-ray Research Institute of the Russian Soviet Federative Socialist Republic, becoming a departmental chief in this Institute three years later. In 1952, she obtained a doctoral degree and was immediately conferred the official state title of professor. Kozlova researched the feasibility of switching the source of radiation from radium to radium-mesothorium or radioactive cobalt. She continually devoted herself to developing radiation therapies for serious malignant tumors. Regarding her teaching activities, in 1954, she became Chief of the Radiation Sickness Department of Central Institute of Advanced Training of Medical Doctors of the USSR Academy of Medical Sciences, and at the same time Dean of the Radiation Therapy Faculty of an evening school attached to the X-ray Research Institute at which she worked as a radiologist. By the age of 60, she had taught approximately 6,000 students.[1] Regarding her activities as a Communist Party member, she served on the Soviet Peace-Protecting Committee.[2] A prolific author, she produced 200 academic articles and ten sole-authored monographs.[3] The Russian State Library lists 16 works that she sole-authored or co-authored between 1939 and 1979, eight of which were produced in a single year, 1955.[4]

1.2 Her involvement in the study of radiation sickness

As mentioned above, Dr. Kozlova was primarily a researcher, physician, and educator in radiotherapy. As such, she was only involved in radiation sickness through her teaching work at the USSR Academy of Medical Sciences. Therefore, of her many publications, only three dealt with radiation sickness.[5] One of these was the pamphlet *Luchevaya Bolezn'* [*Tekst*] (*Radiation Sickness*). Although the date of its publication is unknown, it constitutes one of a series of transcripts of lectures, issued in 1955 by All-Union Society for Promotion of Political and Scientific Knowledge (*Vsesoznoe obshhestvo po rasprostraneniyu politicheskikh i nauchnykh znanii*).[6] The small 24-page booklet has the subtitle *Vliyanie ioniziruyushchego izlucheniya na organizm cheloveka* (*The Effects of Ionizing Radiation on Human Organs*), and about 78,600 copies were distributed to subscribers. The booklet included four photographs, all showing the hands of patients with radiation sickness. No information about the sources was provided for any of the photos. They did have captions, however. The first caption (p. 18) reads, "Tumor that formed in the engineer's hand while preparing radioactive materials"; the second (p. 19) reads, "Skin ulceration in the left ring finger formed after X-ray exposure"; the third (p. 20) reads, "Severe radiation scald caused by a nuclear reactor incident"; and the fourth (p. 20) reads, "Hand several days after a blister was excised." While there is no information regarding the country where the photos were taken, in view of the caption contents, it is difficult to imagine that the persons photographed were victims of the Hiroshima, Nagasaki, or Bikini Atoll blasts. It is far more likely that they were victims of incidents at Soviet nuclear facilities. Indeed, the text as a whole seems to be focused on radiation exposure incidents in the Soviet Union. It was in June 1955 that Kozlova received data on the Hiroshima, Nagasaki, and Bikini Atoll victims, while attending the International Scientific Round-Table on Radiation Effects. Nevertheless, in preparing the above-mentioned work, she did not use such data (leaving aside the oral presentations at the conference that she might have heard). The booklet consists of five chapters: "The units for measuring ionizing radiation and the activity of radioactive material," "The biological effects of ionizing radiation," "The chronic conditions of radiation sickness," "Chronic radiation scalds in the hand," and "Ways to protect against the harmful effects of ionizing radiation."

At the general assembly of the USSR Academy of Sciences held in July 1954, immediately after the first operation of the Obninsk Atomic Power Plant, President (1951–1961) Aleksandr Nesmeyanov (1899–1980) expressed his optimism about the utility of radioactive isotopes, saying, "The nuclear industry will provide science and technology with radioactive material for use in medicine for therapy and diagnosis, in public health, food processing, automatic devices, mineral exploration, and various other areas."[7] Kozlova's booklet was also so optimistic that it extolled the virtues of X-ray testing in the metal industry, the application of radiation in geological surveys for water and oil deposits, radioactive tracing to explore the nutrient intake process in crop and livestock farming, biochemical and physiological applications of the results of tracing and, above all, the use of radiotherapy in clinical practice. For example, Kozlova stated, "It is well accepted that radioactive material has manifold utility in engineering, biology, medicine, agriculture, and various fields of science" (p. 3). On safety standards, which constitute the essential issue, she said here,

> In the case of instant exposure of a whole body, transient and notable changes would appear to living organs at the level of 15 to 20 R (Roentgen) of radiation dose. Slight acute radiation sickness will occur at 100 R; the sicknesses will turn to genuine radiation sicknesses at 400 R, and 50% of sick persons will die. A radiation dose of 600–700 R is regarded as a lethal dose.
>
> (pp. 9–10)

In those days, the Soviet Union generally followed the radiation protection standard set by the International Congress of Radiology (ICR), which was held in London in 1950 (strictly speaking, the standards were recommended by the International Commission on Radiological Protection, which was established and affiliated to the ICR and set up in the same year).[8] We can therefore conclude that Kozlova's recognition of how the effects varied by dose was likely obtained from sources such as the 1950 U.S. report titled *The Effects of Atomic Weapons*,[9] which will be discussed later. Thus, having followed the Western approach to evaluating the effects of radiation, Kozlova concluded as follows:

> Working with radioactive material and ionizing radiation is hazardous. However, the dangers will be considerably minimized if there is a proper labor organization and if workers keep to the codes for adequate radiation protection thoroughly. Optimum protection will be achieved when, in addition to the above, those working near ionizing radiation clearly understand the actual dangers and the available means to protect against it.
>
> (p. 24)

Immediately after the International Scientific Round-Table on Radiation Effects, which took place in Japan in 1955, Kozlova was dispatched to Geneva to attend the First United Nations International Conference on Peaceful Uses of Atomic Energy. At this conference, she gave a presentation entitled "Several Cases Concerning the Therapeutic Application of Radioactive Isotopes."[10] In this presentation, she reported the results of the application of radioactive cobalt, phosphor, iodine, and gold in cancer treatment. Here, she was marking a return to her original field of research.

2 The International Scientific Round-Table on Radiation Effects

In March 1954, the crew of the Japanese fishing boat *Daigo Fukuryū Maru* (*F/V Lucky Dragon 5*) was exposed to and contaminated by fallout from a hydrogen bomb test at Bikini Atoll. The entire crew suffered acute radiation syndrome, and one crewmember later died. After the Bikini Atoll incident, the U.S. government dispatched scientists in and around the U.S. Atomic Energy Commission to Japan to hold confidential talks with Japanese experts; the Japan-U.S. Conference on the Effects and Use of Radiation (Japan-U.S. Radiation Conference) was held behind closed doors on November 15–18.[11] Quite not a few Japanese people have felt suspicious and resentful of the way scientists connected to the government that caused the incident were meeting behind closed doors with a few "handpicked" Japanese scientists in order to determine the scale of the harm caused by the radiation and what verification standards to use.

In July 1954, the International Medical Association for the Study of Present Conditions and Health (AMIEV: an abbreviation of capital letters of French words standing for this organization), a leftwing organization then headquartered in Vienna, sent a letter to Japanese physicians requesting permission to send to Japan an investigating mission composed of medical experts specialized in assessing casualties caused by A and H bombs. Welcoming this offer, the Japanese Association of Medical Doctors against Nuclear Weapons collected signatures and organized a fund-raising campaign for the invitation. In response to the AMIEV's proposal, the World Society on Radiation Medicine also appealed to the scientific societies on radioactivity of each member state to organize arrangements committees in order to hold the World Conference on Radioactivity in Japan together with the AMIEV mission. Responding to the call, a group of volunteers who happened to attend the 14th General Assembly of the Japan Medical Science Society in Kyoto organized the "Executive Committee for the World Conference on Radioactivity (provisionally named)" as the arrangements committee in Japan. The Chairman of the committee was the authoritative and then well-known radiologist Masao Tsuzuki (1892–1961. University of Tokyo); the secretary-general was Hideo Kumabe (1905–1964. Japan Anti-Tuberculosis Association), and the director of the secretariat was Hidetoshi Shiga (1902–?. Institute of Public Health), who was a key figure in the Japanese Association of Medical Doctors against Nuclear Weapons and was closely involved with the All-Japan Federation of Democratic Medical Institutions. In spite of some severe difficulties, especially concerning the immigration hurdle for delegates from the Soviet Union and the People's Republic of China, with which Japan had not yet established normal diplomatic relations, the committee garnered widespread public support and succeeded in holding a major international forum under its new name, the International Scientific Round-Table on Radiation Effects, from May 30 to June 11, 1955.[12]

The opening address at the round-table was delivered by Masao Tsuzuki. As a most "hospitable service" to the delegates from each country, Tsuzuki handed them the data on the casualties in Hiroshima, Nagasaki, and Bikini Atoll. Then, during the round-table, "there were," according to Yasuo Miyake (1908–1990), "reports on the effects of exposure to nuclear explosions, radiation sickness, and the Bikini Atoll incident, and even reports on environmental radioactivity, followed by an intense debate by the foreign scientists."[13]

Here, it is noteworthy that Kozlova was dispatched by the AMIEV to Japan for the round-table.

3 The fate of the data on the Hiroshima, Nagasaki, and Bikini Atoll victims

In 1957, two years after the International Scientific Round-Table on Radiation Effects, Dr. Kozlova translated the data she received at the round-table into Russian and published it in a report titled *The Consequences of Exposure to the Atomic Bombings of Hiroshima and Nagasaki, and the Hydrogen Bombing at Bikini Atoll.*[14] This report consisted of six chapters: "Damage caused by an atomic bomb explosion," "Radiation sickness," "Patients with acute radiation sickness and subsequent changes," "The consequences of the hydrogen bomb explosion at Bikini Atoll (treatment of radiation injuries, metabolism of fission products)," "Radiation in the environment in Japan after the hydrogen bomb explosion at Bikini Atoll," and one other. The first half of the report, concerning Hiroshima and Nagasaki, conveys the findings of a survey conducted by Japanese scientists and physicians in Hiroshima and Nagasaki shortly after the bombings in August 1945 and authorized by the Special Commission for the Atomic Bomb Casualty Survey organized on September 14 (chaired by Haruo Hayashi, Professor Emeritus of University of Tokyo) under the Scientific Research Council of the Ministry of Education, which was later taken over by the newly established Japan Scientific Council.[15] Although no data source was provided for the second half of this report – that is, the final two chapters concerning the Bikini Incident – it is apparent from the content that the report refers to the results of the survey conducted aboard the research vessel *Shunkotsu Maru.*[16] At the end of the report is a five-page-long reference section (pp. 162–166) listing 79 English-language documents and five English-language reports by Japanese authors.

Mikhail Pobedinskii, though reportedly specializing in gynecology, a leading figure in the Soviet medical scientific community in those days, wasted no time in utilizing the above-mentioned data in his popular guide for the public. Entitled *Radiation Sickness: The Consequences of Biological Organ Damage Caused by Nuclear Explosion*,[17] his book contains 60 pages with a rich array of figures and photographs, including photomicrographs of the blood of the *hibakusha*, images of a mushroom cloud, and a photograph of a female *hibakusha* with burnt skin indicating the shadow of a splashed pattern from her clothes. In total, 100,000 copies of this guide were released. After remarking in the introduction that "injuries caused by an atomic bomb explosion will be accompanied by acute forms of light or severe radiation sickness," in a chapter titled "Compound disorders after a nuclear explosion and the treatments thereof" (pp. 47–53), Pobedinskii provides a detailed account of emergent medical treatment in the immediate aftermath of a nuclear bomb explosion on the assumption of civil defense in the case of a possible nuclear war. However, mindful of the recent launch of the Obninsk Atomic Power Plant, which seemingly heralded a new age of the "peaceful use" of atomic energy, Pobedinskii expected the advent of an age in which large amounts of radioactive materials would be used for peaceful purposes, warning,

> If protective measures are not properly adopted, those working around penetrative radiation sources and patients who receive X-ray/radium-based treatment will usually show the signs of chronic forms of radiation sickness. While cases of acute radiation sickness are rare in the peaceful radiation sources, they can occur if the care taken is insufficient.
>
> (pp. 3–4)

How did Dr. Kozlova regard the data on radiation casualties provided by Tsuzuki as his most "hospitable service"? For her, a medical scientist specializing in the medical usage of radiation, the data starkly illustrating the terrible sights of radiation exposure would not have been of primary concern, or at least not her central scientific concern. In the introduction to the Russian version of the report *The Consequences of Exposure to the Atomic Bombings of Hiroshima and Nagasaki, and the Hydrogen Bombing at Bikini Atoll*, she, as a medical specialist involved in radiotherapy, could give no positive evaluation to the report, writing as follows: "Although there has been much valuable reportage on the results of radiation sickness treatment and atomic bomb explosions, this report has shed little light on the many questions that are of interest to the medical community."[18]

In the introduction to a book that Kozlova co-authored with Evgenii Vorob'yov (1918–2007) the following year, *Diagnosis and Treatment of Injuries Caused by a Nuclear Explosion*,[19] the authors state the purpose of the book as follows:

> The purpose of this book is to provide details for the ordinary public on basic matters such as the harmful effects of a nuclear explosion, the schema of treatments for these injuries, basic methods of treating victims of a nuclear disaster, and the principles for protecting against detonation waves, light rays, and ionizing radiation produced in a nuclear explosion.
>
> (p. 3)

They then call on readers to respond calmly in the possible case of a nuclear attack, which then constituted a real and present danger, saying, "Atomic weapons are, like other weapons, not absolute. The effects of injuries caused by a nuclear explosion can be considerably mitigated by taking proper protective measures and rescuing victims in a timely manner" (p. 3). In a reference section two pages long, provided by the authors at the end of the book, the data Tsuzuki handed to Kozlova are unmentioned. Having mentioned the severity of detonation waves and light rays, the strength of radiation, the difference in its effects according to distance from the epicenter, and residual radiation in Hiroshima and Nagasaki, and having cited a Japanese source, *Atomic Bomb Injuries* (1953) written by "Kusano, H." [*sic*; this should be "Kusano, N.," assuming it is Nobuo Kusano], Kozlova and Vorob'yov depended mostly upon the Russian version of *The Effects of Atomic Weapons*, published by the Foreign Literature Publisher in 1954. The original English version (referenced earlier in note 14) was edited by Ashley W. Oughterson and Shields Warren.[20]According to Hiroko Takahashi (2008), Oughterson and Warren were both colonels in the U.S. Army Medical Corps, and from the very early days of the U.S. occupation of Japan, they took part in surveys of the *hibakusha* and the residual radiation in Hiroshima and Nagasaki.[21]

Although the All-Union Scientific and Technical Conference on the Applications of Radioactive and Non-Radioactive Isotopes in National Economy and Scientific Research, held in 1957, deserves our attention for other reasons, featuring as it did presentations of important and original research, it was in her paper in the proceedings of this conference that Kozlova outlined her own optimistic views on the utility of radioactive isotopes. First, however, she admits that there were worries over potential harm from radiation, saying, "In foreign countries, increasing interest in the potentially horrendous consequences of using isotopes has recently been highlighted. Public opinion warning of the carcinogenic effects of radiation is now in ferment. We must investigate this extremely important issue" (p. 3). However, she goes on to argue that "since it will take more than 10 to 20 years for the carcinogenic effects of radiation in humans to become apparent and we have been using radioactive isotopes for no more than 10 years, at this moment we have no way of either denying or confirming this danger" (p. 3). In addition to this remark, she says, "We should have nothing to fear, contrary to some cases in foreign countries, when we evaluate this research method [the use of radioactive isotopes in diagnosis] in the Soviet Union, regarding the above danger." She uses the rest of her paper to demonstrate the utility of radiation in treating malignant tumors.[22]

Notes

1 "Anna Vasil'evna Kozlova (K 60-letiyu so dnya rozhdeniya)." *Meditsinskaya Radiologiya*, Vol. XII, No. 3, 1966, 90–91.
2 "Anna Vasil'evna Kozlova (K 70-letiyu so dnya rozhdeniya)." *Meditsinskaya Radiologiya*, Vol. XXI, No. 9, 1976, 94.
3 "[Nekrolog] Pamyati A.V. Kozlovoi." *Meditsinskaya Radiologiya*, Vol. XXV, No. 7, 1980, 95.
4 The 16 works listed in the Russian State Library are shown below. Please note that most of the works are in a closed collection of that library, meaning that one must visit another library, the Central Scientific Medical Library (*Tsentral'naya nauchno-meditsinskaya biblioteka*), to access them.

(1) A.V. Kozlova, *Rak guby i Ego Lechenie* [*Tekst*]. Tashkent, Gos. izd. nuch.-tekh. i sots.-ekon. lit-ry. UzSSR, 1939 (Seriya monografii/Uzb. In-t eksperim. Meditsiny; Vyp. 1).

(2) A.V. Kozlova, *Luchevaya Terapiya Bol'nykh s Zapushchennykh Formami Zlokachestvennymi Novoobrazovanii* [*Tekst*]. Avtoreferat dis. na soiskanie uchyon. step. d-ra med. nauk. Moskva, Leningr. tsentr rentgenol. radiol. i rakovyi in-t M-va zdravookhraneniya SSSR, 1951.
(3) A.V. Kozlova, *Luchevaya Bolezn'* [*Tekst*]*: Vliyanie Ioniziruyushchego Izlucheniya na Organizm Cheloveka.* Moskva, Znanie, 1955 (Seriya3/ Vsesoyuznoe o-vo po rasprostraneniyu polit. i nauch. Znanii; No. 51).
(4) A.V. Kozlova, "Opyt lechebnogo primeneniya nekotorykh radioaktivnykh izotopov [Tekst]." *Doklady Predstavlennye SSSR na Mezhdnarodnuyu Konferentsiyu po Mirnomu Ispol'zovaniyu Atomnoi Energii (Reports read at the International Conference on Peaceful Use of Atomic Energy).* Moskva, 1955 (Doklady Predstavlennye SSSR na Mezhdnarodnuyu Konferentsiyu po Mirnomu Ispol'zovaniyu Atomnoi Energii.)
(5) A.V. Kozlova, *Zakonomernosti Reaktsii Normal'noi Kozhi i Slizistoi pri Radievoi Terapii Zlokachestvennykh Novoobrazovanii* [*Tekst*]. Moskva, Medgiz, 1955.
(6) A.V. Kozlova, *Lechenie Kapillyarnykh Angiom Radioaktivnym Fosforom* [*Tekst*]. Moskva, Medgiz, 1955.
(7) A.V. Kozlova, and A.S. Pavlov, *Izmeneniya Normal'noi Legoxhnoi Tkani pri Vnutritkanievom Metode Radievoi Terapii* [*Tekst*]. Moskva, Medgiz, 1955.
(8) A.V. Kozlova, and V.G. Ginzburg, *Kliniko-rentgenologicheskie Nablyudeniya nad Rakovym Opukholyami Litsevoi Oblasti pri Radievoi Terapii* [*Tekst*]. Moskva, Medgiz, 1955.
(9) A.V. Kozlova, and Z.G. Ishchenko, *Primenenie Radioaktivnogo Fosfora pri Lechenii Zlokachestvennykh Novoobrazovanii i Predrakovykh Sostoyanii Kozhi i Slizistykh* [*Tekst*]. Moskva, Medgiz, 1955.
(10) A.V. Kozlova, L.G. Fidergoll'ts and Z.F. Lopatnikova, *Lechenie Obshchei Luchevoi Reaktsii pri Radievoi Terapii Opukholei Golody* [*Tekst*]. Moskva, Medgiz, 1955.
(11) A.V. Kozlova, and E.I. Vorov'yov, *Klinika i Lechenie Povrezhdenii Voznikayushchikh pri Vzryve Atomnoi Bomby* [*Tekst*]. Moskva, Medgiz, 1956.
(12) A.V. Kozlova, and G.A. Zubovskii, *Materialy k Lektsii na Teme "Primenenie Radioaktivnykh Izotopov v Meditsine"* [*Tekst*]. Moskva, Vsesoyuznoe obshchestvo po rasprostraneniyu politicheskogo i nauchnogo znanii"(All-Union Society for Diffusion of Political and Scientific Knowledge), 1957.
(13) A.V. Kozlova, V.A. Ankudinov and L.M. Omel'yanov, *Rukovodstvo k Provedeniyu Prakticheskikh Znanii po Klinike Luchevoi Bolezni* [*Tekst*] (Posobie dlya Prepodavatelei). Moskva, Medgiz, 1960.
(14) A.V. Kozlova, [Lektsii] *dlya Vrachei Zaochnogo Tsikla po Klinicheskoi Radiologii* [*Tekst*]. Moskva, M-vo zdravookhraneniya SSSR. Tsentrr. in-t usovershenstvovaniya vrachei, 1961. obl. 1962.
(15) A.V. Kozlova, Yu.Kh. Sarkisiyan and M.M. Khrushchov, *Luchevoe Lechenie Raka Pryamoi Kishki* [*Tekst*]*: (Metod. pis'mo).* Moskva, M-vo zdravookhraneniya RSFSR. Mock. nauch.-issled. Rentgeno-radiol. in-t, 1966.
(16) A.V. Kozlova, V.O. Kalina and Yu.L. Ginzburg, *Opukholi LOR-organov* [*Tekst*]. Moskva, Meditsina, 1979.

5 Those three works were the books 3, 11, and 13 cited in note 9.
6 Kozlova, *Luchevaya Bolezn'* [*Tekst*] . . . *op. cit.*, in note 4. When quoting from this source, the author includes the relevant page numbers in the main text.
7 Arkhiv Rossiskoi Akademii Nauk (ARAN), Fond (F.) 1647, Opis' (Op.) 1, No. 202, 14.
8 A.A. Letaved, "Gigienicheskie problemy v radiologii." Pod red., A.A. Letaveda, *Trudy Vsesoyuznoi Konferentsii po Meditsinskoi Radiologii.* Moskva,

Gosudarstvennoe izdatel'stvo meditsinskoi literatury, 1957, 4. It is noteworthy that an influential and authoritative figure in the Soviet medical scientific community, Avgust Letavet (1893–1984), a full member of the USSR Academy of Medical Sciences and Director of the Institute of Labor Hygiene and Occupational Diseases, highlighting the fluctuation in the Western standards of maximum permissible exposure to radiation during the early 1950s, questioned the arbitrariness of the standards, asking, "Is there any need to further lower the permissible dose of radiation?" He recommended that radiation protection methods be developed on the basis of the Soviet Union's own research and measurements (ibid., 3–10).

9 *The Effects of Atomic Weapons.* Prepared for and in cooperation with the U.S. Department of Defense and the U.S. Atomic Energy Commission under the Direction of the Los Alamos Scientific Laboratory. Los Alamos, New Mexico, June 1950 (http://babel. hathitrust.org/cgi/pt?id=mdp.39015001554743;view=1up;seq=358): Kozlova's understanding of the differentiation of the biological effects of radiation according to differences of dosage, as seen in her booklet, closely resembles the estimations presented in Table 11.28 on page 342 of this book. A Russian version of *The Effects of Atomic Weapons* was published in 1954 (*Deistvie Anomnogo Oruzhiya*, trans. from English. Moscow, Izd-vo Inostrannoi Literatury, 1954): This translated version cites the original text as "Ashley W. Oughterson and Shields Warren, *The Effects of Atomic Weapons.* New York, Toronto, London, 1950," which is slightly discrepant with the title that appears in the original version, as can be seen at the above-mentioned link.

10 The fourth book listed in note 4; Kozlova, *Opyt . . . op. cit.*, in note 4. (13). Another Soviet expert dispatched to Geneva for the conference was Angelina Gus'kova (1924–2015), who later bore important witness to early cases of severe radiation exposure among the workers at nuclear facilities (*Soznanie Pervoi Sovetskoi Yadernoi Bomby.* Moskva, Energoatomizdat, 1995, 148–169; A.K. Gus'kova, *Atomanaya Otrasl' Strany Grazami Vracha.* Moskva, Peal'noe vremya, 2004). Gus'kova gave a presentation, under her own name and that of Grigorij Bajsogolov (1921–2003), on their experiences in treating radiation sickness (ARAN, F. 694, Op. 1, D. 101, 15.). Although Gus'kova is now recognized as one of the leading Soviet figures involved in researching radiation sickness and developing treatments for it, at the time of the conference, she was still only 30 years of age.

11 Yasuo Miyake, *Shi no Hai to Tatakau Kagakusha* (*The Scientists Who Fought Against Radioactive Fallout*). Tokyo, Iwanami Shoten, 1972, 123–125.

12 Shôzô Azami, "Shiga Hidetoshi no Sokuseki (2): Nihon no Gensuikinn Undô tono Kakawari (The Life of Hidetoshi Shiga (2): His Involvement with Japan's Anti-Nuclear Weapons Movement). *Min-iren Iryô* (*Medical Activities of All-Japan Federation of Democratic Medical Institutions*), No. 361, September 2002, 61; *Gensuibaku Kinshi wo Yôbôsuru Ishi no Kai News* (*Japanese Association of Doctors Against Nuclear Weapons News*), No. 5, November 24, 1954." in Tôru Kobayashi (Ed. & Annotation), *Gensuibaku Kinshi Undô Shiryô Shû* (*Materials on the Anti-Nuclear Weapons Movement*), Vol. 1. Tokyo, Ryokuin Shôbô, 1995, 435; Zen-Nihon Minshu Iryô Kikan Rengô Kai (All-Japan Federation of Democratic Medical Institutions (Ed.), *Min-iren Undô no Kiseki* (*Tracing the Progress of the Democratic Medical Institutions Movement*). Tokyo, Ayumi Shuppan, 1983, 164–165. As to the "widespread public support" that the conference garnered, it is noteworthy that then–Prime Minister Ichirô Hatoyama (with his wife Kaoru) and Prince Takahito of Mikasa both donated 5,000 yen to the conference (*Sekai Hôshanô Igaku Kaigi Sokuhô – Bulletin of the World Conference of Radiology*, No. 2, in *Gensuibaku Kinshi Undô . . . op. cit.*, Vol. 2. 109).

13 Miyake, *op. cit.*, 137.

14 A.V. Kozlova (Sost.), *Posledstviya Vzryvov Atomnykh Bomb v Khirosime i Nagasaki i Vodorodnoi Bomby v Bikini [Tekst]: Otchyot ob Intern. konferentsii v Tokio 1955g., Posvyashch. Posledstviyam Vzryvov Atomnoi i Vodorodnoi Bomb.* Moscow, Medgiz, 1957, 168.

15 Ibid., 5–6. It seems that the source of her translation here was an abridged English translation of the original Japanese report, Nihon Gakujutu Kaigi (Japan Scientific Council) (Ed.), *Bakudan Saigai Chôsa Hôkokusho* (*Atomic Bomb Casualty Survey Reports*), in five volumes (summarized edition published in 1951, Vol. 1–5, 1953). However, because she neglected to include certain information, such as the names of the authors of each chapter, it is not possible to trace back the sources to determine which parts were the translations. Also of note, the introduction includes a list of members of the sections of the Special Commission for the Atomic Bomb Casualty Survey (p. 5). However, the surnames of the chairs of these sections there have been transcribed rather sloppily; (Masaichi) Majima was transcribed as "Mayjima," (Shôji) Setô as "Sobo," (Ikusaku) Amemiya as "Amenija," (Ihachirô) Miura as "Miusa," and (Kiyoshi) Masui as "Masao."

16 Ibid., 140–160. Here too Kozlova and other editors do not cite the names of authors or research organizations, which makes the sources of her data unclear. The Japanese data on the Bikini Atoll incident that were published prior to the International Scientific Round-Table on Radiation Effects of June 1955 include the first volume of a report from the Research Division of the Fisheries Agency entitled *Bikini Kai-iki niokeru Hôshanô Eikyô Chôsa Hôkoku* (*Survey Report on the Effects of Radiation in the Bikini Atoll Area*) (summarized version: Miyake, Yasuo, Hiyama, Yoshio, Kusano, Nobuo [supervision], *Daigo Fukuryû Maru* Heiwa Kyôkai (Ed.), *Bikini Suibaku Hisai Shiryô Shû – Collection of Materials on the Bikini Hydrogen Explosion Casualty.* Tokyo, University of Tokyo Press, 1976, 112–119). The part in the Russian version corresponding to this part was probably translated from an abridged English translation of this report.

17 M.N. Pobedinskii, *Luchevaya Bolezn'* (*Posledstviya Porazheniya Organizma pri Atomnom Vzryve*). Moskva, Medgiz, 1957g. Quoting from this book, the author includes the relevant page numbers directly in the main text. It is noteworthy that the book encourages readers to remain tranquil, stating, "A series of measures and methods are used to prevent or treat damages to organs caused by radiation, including measures to restrain the appearance of radiation sickness or extensively mitigate its progress. Further research by the medical doctors who devote themselves to fighting this disease will undoubtedly enable great successes through radiotherapy" (60).

18 Kozlova (Sost.), *Posledstviya vzryvov atomnykh bomb . . . op. cit.*, in note 19. 4.

19 Kozlova and Vorov'yov, *Klinika . . . op. cit.*, in note 4 (item 11). Quoting from this book, the author includes the relevant page numbers directly in the main text.

20 *Deistvie anomnogo oruzhiya . . . op. cit.*, in note 9.

21 Hiroko Takahashi, [Shintei/Zôho Ban] *Hûin sareta Hiroshima, Nagasaki: Bei Kaku-jikkenn to Minkan Bôei Keikaku* (*Hiroshima and Nagasaki Sealed-off: U.S. Nuclear Testing and Civil Defense Plan)* (New and Expanded Edition). Tokyo, Gaifûsha, 2008, 66–68.

22 A.V. Kozlova, "Primenenie radioaktivnykh izotopov v klinicheskoi meditsine." in *Meditsinskaya radiologiya: Trudy Vsesoyuznoi nauchno-tekhnicheskoi konferentsii po primeneniyu radioaktivnykh i stabil'nykh izotopov i izluchenii v narodnom khozyaistve i nauke.* Moskva, Izd-vo AN SSSR, 1960, 3–14.

Conclusion

In the first page of this book, we posed the questions: "Why were such a great number of scientific talents mobilized for various Cold War research and development projects? How can we explain that reason in the light of the newest achievements of the historical study, in which the simple totalitarian model has already lost credence? How can we characterize the basic features of the Soviet Cold War research and development projects in their realities? What were the real consequences of the rushed projects? How did Soviet scientists approach their nuclear age?"

Concerning the first two questions, we inquired into the academic "discussion" campaign conducted in the late 1940s in the field of physics, paying major attention to the formation of the "Cold War climate." Social "climate," including some kind of mass-hysteria, has great significance in countries such as Russia, Japan, and others where modern civil society was developed relatively weak.

Could there have been any genuine physicists who did not know anything about quantum dynamics and the principle of relativity by the second half of the 1940s? As I referred in Part I, Chapter 2, in 1951, a Moscow State University (MSU) teacher of physical chemistry, Sergei Vasil'ev (1904–1999), a supporter of the "patriotic and materialistic physicists" in the postwar academic "discussions," was in turn indicted by the Dean of the Faculty of Chemistry of MSU for his "double-dealing" attitude toward "idealistic" physical theories; even while he outwardly condemned "idealism" in physics, he taught students the "idealistic" theories of Niels Bohr (1885–1962), Ernest Rutherford (1871–1937), Linus Pauling (1901–1994), and Erwin Schrödinger (1887–1961) in his lectures.[1] Furthermore, we must cite the terribly low level of "philosophical" acumen of the "patriotic and materialistic physicists"; for example, in a "philosophical" paper published in an authoritative journal of physics, a staff member of the Faculty of Physics, MSU, Fyodor Koloryov (1909–1979) condemned his colleague Semyon Khaikin for "idealism" for no other reason than his terminology.[2]

During the war, the majority of the institutes of the Soviet Academy of Sciences continued scientific research, but it was mainly intended for military purposes. Significant results were achieved in various fields, but the wartime experience deepened the "rift" between scientists inside and outside the Academy. In particular, the wartime evacuation of the scientific research institutes and the institutions of higher education to different places strengthened the tendency toward a functional separation of the Academy of Sciences and universities, which had already exacerbated a serious confrontation among Soviet scientists even before the war.

The initiation of the apparently ideologically motivated campaign in various fields of science rekindled the jealousy and hatred accumulated on the side of university professors and lecturers toward some of their "colleagues" with a record of splendid academic and scientific achievements. They attempted to challenge these accomplished scientists in spite of their own lack of experience and ability in philosophical thinking and without help from a genuine philosopher or ideologue other than the unfashionable philosopher Aleksandr Maksimov (1891–1976).[3] The frustration of "patriotic and materialistic physicists" accumulated to such an extent that the leading echelon of the Party bureaucrats in charge of scientific matters, so insightfully, recognized the necessity to let such a miasma out of the Soviet academic community, avoiding adroitly "explosion," some possible harmful influences upon scientific research that might have been caused by possible blunt intervention of Iosif Stalin.

The ideological campaign was conducted in its essential parts by neither the philosophers nor the Party ideologues, but by the physicists themselves who called themselves the "patriotic and materialistic physicists." The Party and governmental organs played their role as a mediator in this case. No matter whether the All-Union Physics Conference was held or not, the political leadership intended simply to exclude the miasma undermining the Soviet physics society and to bring about some ideological unification among the physicists. And then, no matter whether it was expected from the initial stage of the campaign or not, the campaign succeeded in the creation of the "Cold War climate" among the physicists which tacitly drove many physicists into the devotion of themselves to several Cold War scientific projects.

Our next questions were, "How can we characterize the basic features of the Soviet Cold War R&D projects in their realities? What were the real consequences of the rushed projects?"

Despite little attention paid to such a fact, there was an "interval" between the victorious end of World War II and the outbreak of the Cold War for only a few years, or only some months in some occasions, in which the disarmament of the extremely enlarged war potential, the conversion of over-expanded military production to the peaceful one, and the dissolution of war economy in general were largely promoted by the Soviet government. During World War II, the important parts of the territory of the Soviet Union turned into battle fields in which as many as 20 million human lives were lost. The Soviet national economy was also heavily damaged; 37 iron and steel metallurgical factories, 1,135 coal mine shafts, 61 large electric power stations, 10,000 km of high-voltage power transmission line, 746 machine-building factories, 50% of nitrogen productive capacity, 77% of sulfuric acid productive capacity, 88% of productive capacity of anhydrous sodium carbonate, and others were destroyed.[4] It is needless to say that the ministries related to military production were also expected to make great contributions to the postwar economic recovery.

For the young, talented bureaucrats who were selected early in their careers to the top management of various industrial people's commissariats – owing to the large vacancy of able management staffs caused by the Great Terror – and made their splendid careers in the war economy, such conversion of the policy line was a peril. They were required to find a new way suitable for the maintenance of their statuses and their career-advancement.

The response of the Ministry of Armament to the rocket development is suggestive about the interests and motivations of the top management of the industrial ministries, which intended to build up large-scale research and development organizations for new military technology.

Trial of Ministry of Armament for conversion to peaceful production, however, ended in failure. The difficulties in the management of the factories affiliated to the Ministry of Armament were so serious that it would have not been so astonishing to imagine the abolishment of the Ministry. This issue was also so serious for the future of Minister Dmitrii Ustinov (1908–1984) himself.[5]

On the other hand, German rocket V-2 was not so promising because of its technological defects such as inaccuracy of control and others. Even if such defects had been overcome, its efficiency would have been disappointing in light of its expensiveness, when it had been loaded with only a conventional warhead. Only in the midst of 1950s, when the rockets were included into nuclear strategy as the transporting means of nuclear warheads, the rockets were recognized to be promising. However, the rocket development was "a straw a drowning man will catch at," or "welcome rain during the dry season" for the Ministry of Armament, irrespective of whether the rocket was promising or not.

In this meaning, Ustinov eventually won a bet, whereas Pyotr Parshin (1899–1970), Minister of Machine-building and Instrument-building (*Ministr Mashinostroeniya i Priborostroeniya*. The name of his Ministry is abbreviated as MMiP), lost his bet. His main client, the Soviet military, especially its artillerist part, was content with the mechanical calculators and the analogue machines provided by his ministry, the MMiP, which was familiar to the military elite, especially the artillerymen, as a supplier of trench mortars, rocket launchers, mines, and shells, and other projectile weapons for them.[6] Moreover, the artillery part of the military seems to have preferred the special-purpose analogue calculators because they were easier to operate and a lot cheaper.[7]

The artillerists were not the primary force behind the desire for faster computers. It seems it was a group of scientists and mathematicians, many of whom engaged themselves in the Soviet nuclear project, such as Mikahil Lavrent'ev (1900–1980), that felt the real need for general-purpose, electronic, digital computers, for example to process advanced and complicated calculations required for nuclear physics. In Soviet bureaucratic culture, in order to eliminate obstacles in the way of computer development, they had to seek patronage from the top political leadership. It seems that the scientists' approach to the political leadership meant, at the same time, their entry into a symbiotic relationship with each other.

Special Design Bureau No. 245 (*Spetsial'noe Konstruktorskoe Byuro No. 245*; SKB-245) of MMiP took countermeasures against Lavrent'ev's group – the development of their own computer. By the means of harassment, they eventually gained momentary success.

All of these processes shaped the Soviet quest for high-speed computers into a complicated and discordant process. Without any doubt, they greatly delayed Soviet computer development.

In many cases, the postwar Soviet military technological project took advantage of the German technological seeds obtained by the Soviets during the occupation. The development of the first jet planes of the Soviet Union involved not only a large number of drawings and design data, but also approximately 8,000 German engineers and workers, among whom, approximately 1,400 were transported to Soviet territory (3,500 people including their families), and totally 65,960 units (in the plans) of machinery and equipment from 30 factories confiscated in Germany. That was, no doubt, a process of "plunder" of technology in an unprecedented scale.

However, despite the attempt at such large-scale technology transfer and strict orders from Stalin, it was difficult to acquire the necessary equipment, raw material, and human resources under the condition of a divided occupation. Beside it, the mutual agreement with other Allied countries on prohibition of the research and development activities in each occupied area of Germany, the lack of good communication among the Soviet military personnel and others compelled the research and development activities in German territory to be transferred to the Soviet territory while still unfinished. Consequently, in a sort of irony, on many occasions, German jet engine technology and Soviet airframe design often had to be "merged" or, to be more precise, complementarily combined as a compromise measure, as indicated in the typical examples of Me-262 and MiG-9.

Undoubtedly, this "successful" jet plane development was a landmark process that elevated Ministry of Aviation Industry as a military industry agency.

Before the war, in 1940, the People's Commissariat of Aviation Industry employed 370,200 workers in exactly 100 factories (approximately 41,000 metal-cutting machine tools), but this number expanded to 171 factories, 93,000 units of equipment, and 715,000 employees at its peak in 1945. In 1946, it decreased to 158 factories and 623,300 people, but only the number of machine tools increased to 106,000 units.[8] The People's Commissariat of Aviation Industry hurriedly recruited 218,000 people in December 1941 (among whom, 137,000 were licensed engineers) for the wartime evacuation of the factories to the eastern part of the territory. Of the total ammunition-industry-related recruitment of 470,000 jobs, this was the largest number recruited by a single commissariat (People's Commissariat of Ammunition, 136,000 people; People's Commissariat of Armament, 64,200 people; People's Commissariat of Tanks, 45,000 people).[9]

In accordance with the postwar "civilian conversion," the People's Commissariat of Aviation Industry also increased the supply of energy equipment, electric power equipment, buses, and motorcycles. However, in its fifth five-year plan from 1946 to 1950, 25,765 military aircrafts (out of which, 5,700 were jet planes)[10] were planned to be produced – a sign that the commissariat was returning to its nature of military industry agency. In the plan for 1946, the production of civilian goods accounted for 50.2% of the Ministry of Aviation Industry and this number in the plan for 1950 was 46.8%, but the actual result of the same year was 25.1% (in other words, 74.9% were military products).[11] In October 1952, immediately before the Nineteenth Congress of the Communist Party, in "Material concerning to the development of the Soviet machine-building industry between 1938 and 1950," prepared by the Machine-building Industry Division of the Party's Central Committee, Ministry of Aviation Industry, during this period, produced 20 types of airplanes (out of which, nine were in serial production) and developed 30 types of engines and jet engines (two types of jet engines were transferred to serial production). In 1946, jet planes accounted for only 1% of the total production of aircrafts,[12] but in 1948, of the 6,530 planes produced, 750 were jet planes.[13] Subsequently, in 1950, this number had already reached 65%.[14]

Our last question was, "How did Soviet scientists approach their nuclear age?"

The first nuclear power plant in the world went into operation at Obninsk, a southern suburb of Moscow, in July 1954. One year later, the Soviet Academy of Sciences held a large-scale international conference, the Session of the Academy of Sciences of the USSR on the Peaceful Use of Atomic Energy, as an international stage to demonstrate their scientific achievements in nuclear science and technology. Soviet scientists subsequently challenged the United States' nuclear advancement at the First United Nations International Conference on the Peaceful Use of Atomic Energy at Geneva. Could the Soviet scientists succeed in gaining international acceptance and prestige? Could they gain self-confidence?

Heinz Barwich (1911–1966. Later, he immigrated to Western Germany in 1964), the first Director of the Central Institute for Nuclear Physics (Zentralinstitut für Kernphysik) in the German Democratic Republic (GDR), was regarded as "a friend of the Soviet Union" and as a sympathizer of Soviet socialism who had spent ten years in the Soviet Union and preferred the "socialist" style of talks. However, he did not choose a graphite-moderator type of reactor for the first research reactor in the GDR; instead, he chose a light-water moderator-coolant type reactor. He was proud of "the fact that our research reactor differs substantially in a number of characteristics from the reactor model which was already taken into operation 15 years ago by American researchers and recently taken into operation by Soviet researchers for the first time."[15] We must take into consideration that the development of a light-water reactor was still a big challenge at that time. What is more, the GDR aimed to construct heavy water reactors, and then, fast-breeder reactors as a main method of settlement of the energy issues in those days.[16] In other words, the graphite reactor was already not very attractive and somewhat out-of-date for the GDR ambitious scientists. Even "friends of the Soviet Union," like Barwich, could not have been very satisfied with the small graphite-moderator reactor at the Obninsk Atomic Power Station, which Barwich observed at the actual site.

The absence of a number of eminent scientists due to the poor arrangement by the Academy of Sciences and other Soviet agencies and the attendance of some not very earnest participants who were not so interested in the achievements of Soviet science made the Session unsuccessful. In addition to this, the Session provided an occasion for some Western scientists who were skeptical or critical of the Soviet Union to take advantage of the unguarded openness of the Soviet scientists. Sigvard Eklund (1911–2000. Later, the Director of the International Atomic Energy Agency, 1961–1981) and others gathering information in Moscow was another embarrassment.

At the Geneva Conference, the U.S. delegation overwhelmed Soviet scientists, and "the atomic diplomacy" practiced by Soviet scientists thus came to a bitter end.

This book has shown that, even before the fierce debate began in the United Nations Scientific Committee on the Effects of Atomic Radiation (UNSCEAR) in the summer of 1958, there had already been an internal drive within the Soviet Union for expanding research on radiation effects on humans and other organisms. Some Soviet scientists and medical scientists had been involved in the disposal of radioactive wastes, the rescue and treatment of persons exposed to radiation, the development of measurement methods and radiation protection, and other activities in Chelyabinsk-40, where a series of radiation exposure incidents repeatedly occurred, and also in relation to the East Ural radiological disaster in September 1957. Taking the opportunity to raise the issue along with global fallout, the research results accumulated by the Soviet scientists thus appeared on the stage of international politics.

From the very early days of research activities on radiation effects in the Soviet Union, there emerged some unique developments. There was scientific interest in a wide range of topics that few Western scientists investigated in a systematic way, such as internal exposure, radiation effects on the nerve activities of animals, and external damage to the tissue caused by some unpenetrated radioactive rays. Well before 1955, when the Soviet Academy of Sciences held the "Session for the Peaceful Uses of the Atomic Energy" in conjunction with the beginning of peaceful uses of atomic energy, biological effects of radiation already had been considered in the Soviet Union as an independent, important area of scientific inquiry. In addition, the 1955 Session saw some serious critiques leveled against the kind of risk analysis practiced in the West as ignorance or underestimation of internal exposure, difference in radiation effects due to diversity in diet and the formation of radioactive carbon in the air and its effects, traps of the "averaging method" and others.

Following the heated debates in UNSCEAR in the summer of 1958, Andrei Lebedinskii (1902–1965), the Soviet chief delegate to the committee, became active in publishing counterarguments against the Western view on radiation effects in the international journal *Atomnaya Energiya* (*Atomic Energy*).[17] Ovsei Leipunskii (1909–1990) and Andrei Sakharov (1921–1989) also contributed to that journal.[18] In this way, the Soviet scientists rapidly prepared their major counterarguments against the prevailing ideas in the West regarding biological effects of radiation. These radical critiques crystalized into the 1959 book, A.V. Lebedinskii (ed.), *Sovetskie Uchyonye ob Opasnosti Ispytanii Yadernogo Oruzhiya* (Moscow, Atomizdat. 1959; *Soviet Scientists on the Danger of Nuclear Weapon Tests*). As has been seen in this book, however, these counterarguments, at least in part, had been prepared in the indigenous research tradition in the Soviet Union based on their own experiences with radiation hazards.

It must be noted, however, that another current – a very optimistic view toward radiation hazards – also emerged in the Soviet Union as early as by the end of the

1950s, suggesting that not all Soviet scientists in those days agreed with radical criticisms leveled against Western studies on radiation effects. As is well known, such optimism later became mainstream even in the Soviet Union.

This book has also demonstrated that the Lysenkoites' opposition was directed not to radiation studies per se but to the foundation of the new research centers outside their sphere of influence for fear that these institutions might underline their monopoly in the biological sciences. Zhores Medvedev[19] connects the Lysenkoites' hegemony with the history of the Soviet agriculture. Such a point of view is, no doubt, reasonable. It must be, however, also considered that the postwar Soviet Union was directly involved in the Cold War and Soviet scientists were required to make contributions to the preparations for a possible war.[20] In such circumstances, a "monopoly of scientific truth" by Lysenkoites must have been trivial. Lysenkoites themselves must have realized this well. It is noteworthy that although Lysenkoites harassed other scientists with their persistent attacks against the foundation of another scientific center independent from their research centers in order to maintain their dominance in biological sciences, they did not oppose the development of radiation effect studies. Ironically they tried their entry into the biological study on radiation effects and justified their interest and interference in the name of "pluralism in scientific research" to shield their genuine intentions.

The "Bikini Incident," a large-scale radiation exposure event caused by a U.S. hydrogen bomb test conducted in the Pacific Ocean in March 1954, shocked the world and, in the second half of the 1950s, there brought about a serious debate on radiation's effects on living bodies in the international arena, such as UNSCEAR. In Japan, one reaction to the "Bikini Incident" was a scientific conference, the International Scientific Round-Table on Radiation Effects, which was held in Japan in 1955 on the basis of support from medical doctors and citizens who were skeptical about the U.S. method of estimating the damage from the "Bikini Incident," which dominated the discussions at the Japan-U.S. Conference on the Effects and Use of Radiation in November 1954. The promoters of that round-table eventually invited nine overseas guests from nine countries, including the People's Republic of China and the Soviet Union, which had not yet established diplomatic relations with Japan. The guests received data on the radiation exposure from Hiroshima, Nagasaki, and Bikini from Masao Tsuzuki (1892–1961), president of the executive committee of the round-table. A medical doctor from the Soviet Union, Anna Kozlova (1906–1980), was one of the guests, and she brought the data back to the Soviet Union.

The data Kozlova brought back with her from the International Scientific Round-Table on Radiation Effects in 1955 did attract the practical interest of medical scientists and were even used in several enlightening books for the public.

However, after translating and publishing the data on the casualties in Hiroshima, Nagasaki, and Bikini Atoll that Tsuzuki had provided her, rather than use the data directly in her own research or in enlightening activities, Kozlova returned to her original field of interest, research in radiotherapy, where she would aggressively promote the diagnostic application of radiation.[21]

Although she would continue to be an active author thereafter, she would never mention anything to do with injuries or radiation sickness caused by nuclear bomb explosions. Evidently, the data handed to her by Tsuzuki as the most "hospitable gift" must have been seen by her as something of an unwelcome favor – a tiresome business to get free of as soon as possible. After the publication of the Russian version of the data, there is no trace of her ever mentioning it again.[22]

In 1973, Koslova contributed an article to *Vestnik Rentgenologii i Radiologii* (*Bulletin of Roentgenology and Radiology*) to commemorate the 75th anniversary of the discovery of radium.[23] There she praised the great significance of the discovery of radium and its medical application and other usages.

It was a "misfortune" for the organizers of that International Scientific Round-Table that the data on the radiation exposure from Hiroshima, Nagasaki, and Bikini were handed to such a person who had such a peculiar, extremely optimistic view on the usefulness of ionizing radiation. Contrary to the expectation of Tsuzuki and other Japanese persons concerned, Kozlova herself was solely interested in the medical applications of radiation and tended to take an optimistic line regarding the potential dangers of radiation.

Notes

1 RGASPI, F. 17, Op. 125, D. 618, 176.
2 See, F.A. Korolyov, "O metodorogicheskikh oshibakh v Knige prof. S.E. Khaikina Mekhanika." *Uspekhi Fizicheskikh Nauk*, Vol. 37, No. 3, March 1949, 388–394; This paper shows that the "patriotic and materialistic physicists" were not qualified to conduct philosophical debates. However, an MSU professor, Yakov Terletskii (1912–1993), is an interesting exception; he devoted himself so enthusiastically to the philosophical interpretation of contemporary physics in search of "physics" consistent with Marxist philosophy. See, a series of his philosophical articles published in the journal, *Voprosy Filosofii*; "Problemy razvitiya kvantvoi teorii" (No. 5, 1951, 51–61), "O soderzhanii sovremennoi fizicheskoi teorii prostanstva i bremeni" (No. 3, 1952, 191–197), "Ob izlozhenii osnov spetsial'noi teorii otnositel'nosti" (No. 4, 1953, 207–212), "O vzaimoprevrashchaemosti elemental'nykh chastits" (No. 2, 1956), etc.
3 Maksimov published his article "Ob odnom filosofskom Kentavre (On One Philosophical Centaur) " in the *Literaturnaya Gazeta* of April 10, 1948. He could not gain the approval of any physicists, however, except for Terletskii (A.S. Sonin, *"Fizicheskii Idealizm": Istoria Odnoi Ideologicheskoi Kampanii*. Moscow, Fiziko-Matmaticheskaya Literatura, 1994, 90–93).
4 I.M. Brover, *Ocherki Razvitiya Tyazheloi Promyshrennosti SSSR*. Alma-A-ta (now, Almaty), AN Kazakhskoi SSR, 1957, 214.
5 Ustinov, an assembler in a factory, became an engineer after graduation from an institute and began to work in a military-related factory, "Bol'shevik" in Leningrad. Only one year after, he was promoted to the Director of the factory, because almost all senior engineers and administrators were arrested in the violent stream of Great Terror. Eventually he was appointed a people's commissar at the age of only 33 in the occasion of reorganization of the people's commissariats related with the outbreak of the war (K.A. Zalesskii, *Imperiya Stalina: Biograficheskii Entsiklopedicheskii Slovar'*. Moscow, VEChE, 2000, 455).

6 This Ministry, then People's Commissariat, was temporarily reorganized as the People's Commissariat for Projectile Munitions, during the war (N.S. Simonov, *Voenno-Promyshlennyi Kompleks SSSR v 1920–1950-e gody: Tempy Ekonomicheskogo Rosta, Struktura, Organizatsiya Proizvodstva i Upravlenie.* Moscow, ROSSPEN, 1996, 139).

7 The planned price of "Strela" (in September 1954), 9,800 rubles, was much more expensive than those of an analogue calculator, 1,350 rubles, or a special-purpose computing machine, 1,000–3,000 rubles (RGAE, F. 8123, Op. 8, D. 551, 29): As well as this, we must point out the tendency of delaying installation. The completion of Strelas was delayed in many cases. For example, the machines for the Ministry of Medium-size Machine-building and Moscow State University were installed only in 1955, almost one year after the plan (RGAE, F. 8123, Op. 8, D. 619, 9, 11).

8 Simonov, *op. cit.*, in note 6. 156.

9 Ibid., 174.

10 Ibid., 195, 196.

11 Ibid., 204.

12 Ibid., 202.

13 Ibid., 238.

14 Ibid., 202.

15 Heinz Barwich, "Uber den Forschungsreaktor der DDR und seine Ausnutzungsmoeglichkeiten." in Heinz Barwich et al. (Ed.), *Das zenralinstitut fur Kernphysik am Beginn seiner Arbeit.* Berlin, Akademie-Verklag, 1958, 8.

16 Mike Reichert, *Kernenergiewirtschaft in der DDR. Entwicklungsbedingungen, konzeptioneller Anspruch und Realisierungsgrad (1955–1990).* Gutenberg, St. Katharinen, 1999, 153–176.

17 He published an article in cooperation with Yu.G. Grigoir'ev and G.G. Dimirchoglyan, "O biologicheskom deistvii ioniziruyushchego izlucheniya v malom doze (On the Biological Effects of Ionizing Radiation in the Small Doses)" in the journal, *Atomnaya Energiya* (*Atomic Energy*), Vol. 5, No. 3, 1958, 310–320 and his sole article, "O biologicheskom deistvii izluchenii (On the Biological Effects of Radiation)." *Atomnaya Energiya* (*Atomic Energy*), Vol. 6, No. 2, 1959, 187–199.

18 Ovsei Leipunskii contributed his sole article, "O radioaktivnoi opasnosti nepreryvnykh ispytanii atomnykh bomb (On the Radioacitive Danger of the Continuous A-Bomb Tests)" to the journal, *Atomnaya Energiya* (*Atomic Energy*), Vol. 4, No. 1, 1958, 63–70. Sakharov wrote his article, "Radioaktivnyi uglerod yadernykh vzryvov i neporogovye biologicheskie effekty (Radioactive Carbon by Nuclear Explosions and Its Non-Threshold Biological Effects)." *Atomnaya Energiya* (*Atomic Energy*), Vol. 4, No. 6, 1958, 576–580.

19 See, Zhores A. Medvedev, *The Rise and Fall of T.D. Lysenko.* New York, Colombia University Press, 1969.

20 At the same time, the leading scientists were also eager to secure "freedom of research" from the political authority and tried to bind these two aspects together. At the above-mentioned Presidium meeting on December 6, 1957, Kapitsa also said, "And so it seemed to me that the experience at Cambridge, Oxford, and other universities where researchers are able to speak freely is needed to apply so as to solve such problems [study on radiation effects – *Ichikawa*]. I think that we can organize a club where people gather and talk, however, it [the solution of the problem – *Ichikawa*] can be done only when people freely exchange opinions. We do not have such free communication. We are closed in their institutions, and see others only at meetings, where if you can talk, the speakers usually interfere (ARAN, F. 2, Op. 6, No. 264, 132)."

21 The journals *Meditsinskaya Radiologiya* (*Medical Radiation*) and *Vestnik Rentgenologii i Radiologii* featured various articles sole-authored or co-authored by Kozlova, but all of these articles concerned the research and development of radiotherapy techniques; none dealt with radiation sickness. That said, toward the end of her highly productive life in research, Kozlova penned an article entitled "How the Results of Radiotherapy for Some Types of Malignant Tumor Depend on the Differences of a One-time Dose of Radiation" (A.V. Kozlova, "Rezul'taty luchevoi terapii nekotorykh form zlokachestvennykh opukholei v zavisimosti ot verichiny razovoi dozy." *Meditsinskaya Radiologiya*, Vol. XV, No. 10, 19–25.) This article apparently represents the essence of her study of radiation dosages as part of her research in therapeutic radiation.

22 *Meditsinskaya Radiologiya*, a journal for which Kozlova served for many years, as an editorial board member and then as chief editor, carried articles commemorating her 60th and 70th birthdays, as well as an article commemorating her passing in 1980 (articles cited in notes 6–8). None of these articles mention her encounter with the data on nuclear casualties in Hiroshima, Nagasaki, or Bikini Atoll, nor do they mention her visit to Japan. However, 20 years after her death, the successor to the above-mentioned journal ran an article commemorating the 100th anniversary of her birth, in which it made a short reference to her visit to Japan ("Pamyati Anny Vasil'evny Kozlovoi. K 100-letiyu so dnya rozhdeniya (1906–1980)." *Meditsinskaya radiologiya i radiatsionnaya bezopasnost'*, Vol. 52, No. 1, 2006, 76–77). This article mentions that she was dispatched to Hiroshima and Nagasaki, representing the Soviet Union, to investigate the real damage caused by an A-bomb there, that she paid her respects to the victims of both cities by laying a floral wreath, writing on the tape of the wreath, "May this error never be repeated," and that she continued to give assistance to Japanese residents suffering from radiation sickness even after her return (she was a member of the board of directors of the Japan-Soviet Friendship Association). Is it telling that the above-mentioned facts only appeared in the journal after the breakup of the Soviet Union and the journal's name change from *Meditsinskaya Radiologiya* (*Medical Radiology*) to *Meditsinskaya Radiologiya i Radiatsionnaya Bezopasnost* (*Medical Radiology and Radiation Safety*)?

23 A.V. Kozlova, "75-let so vremeni otkrytiya radiya." *Vestnik Rentgenologii i Radiologii*, No. 5, 1973, 97–99: In this article, assessing the discovery of radium as equal in importance to that of radioactivity by Antoine Henri Becquerel (1852–1908), she writes, "(The discovery of radium) has enabled humankind to come to the Atomic Age, an age of scientific revolution, and to achieve the greatest works in every field of knowledge, from the composition of matter to the evolution of the cosmos" (97). She then praises the great feat of Marie Skłodowska Curie (1867–1934), saying, "Let us remember the forerunner who, through her discovery 75 years ago, presaged this new age, which can in fairness be called the Atomic Century" (99). In this account, she outlined in detail Curie's deeds in World War I, particularly how she evacuated the precious radium (her X-ray source), organized a fleet of vehicles equipped with X-ray devices, and how she rode with her daughter Irène in an "X-ray vehicle" to the front for the medical examination of injured troops. Kozlova clearly saw in Curie something of herself, as someone who sought to use radiation to liberate humankind from malignant neoplasms.

Name index

Subject index

For Product Safety Concerns and Information please contact our EU representative GPSR@taylorandfrancis.com
Taylor & Francis Verlag GmbH, Kaufingerstraße 24, 80331 München, Germany

www.ingramcontent.com/pod-product-compliance
Ingram Content Group UK Ltd.
Pitfield, Milton Keynes, MK11 3LW, UK
UKHW011012180926
475701UK00004B/699

* 9 7 8 0 3 6 7 5 8 5 2 2 8 *